DECISION-MAKERS FIELD CONFERENCE 2007

WATER RESOURCES OF THE MIDDLE RIO GRANDE

San Acacia to Elephant Butte

L. Greer Price, Peggy S. Johnson, and Douglas Bland, Editors

New Mexico Bureau of Geology and Mineral Resources
A DIVISION OF NEW MEXICO INSTITUTE OF MINING AND TECHNOLOGY
2007

Water Resources of the Middle Rio Grande: San Acacia to Elephant Butte
L. Greer Price, Peggy S. Johnson, and Douglas Bland, Editors

Copyright © 2007
New Mexico Bureau of Geology and Mineral Resources
Peter A. Scholle, *Director and State Geologist*

a division of

New Mexico Institute of Mining and Technology
Daniel H. López, *President*

BOARD OF REGENTS

Ex-Officio
Bill Richardson, Governor of New Mexico
Beverlee J. McClure, Secretary of Higher Education

Appointed
Jerry A. Armijo, 2003–2009, Socorro
Richard N. Carpenter, 2003–2009, Santa Fe
Ann Murphy Daily, 2005-2011, Santa Fe
Abe Silver, Jr., 2007–2013, Santa Fe
Dennise Trujillo, 2007–2009, Socorro (Regent designate)

DESIGN & LAYOUT: Gina D'Ambrosio
SERIES DESIGN: Christina Watkins

EDITING: Jane C. Love and Nancy S. Gilson
SERIES EDITOR: L. Greer Price

CARTOGRAPHY & GRAPHICS: Leo Gabaldon and Tom Kaus
CARTOGRAPHIC SUPPORT: Kathryn Glesener, Brigitte Felix Kludt, and Glen Jones
EDITORIAL ASSISTANCE: Lynne Kurilovitch

New Mexico Bureau of Geology and Mineral Resources
801 Leroy Place
Socorro, NM 87801-4796
(505) 835-5420
http://geoinfo.nmt.edu

ISBN 978-1-883905-24-8
First Printing May 2007

COVER PHOTO:
The Rio Grande at the Bosque del Apache National Wildlife Refuge
Copyright © Adriel Heisey

**The New Mexico Bureau of Geology and Mineral Resources
wishes to thank the following for their support
of this year's conference and guidebook:**

SUPPORTING ORGANIZATIONS

New Mexico State Legislature
New Mexico Office of the State Engineer/Interstate Stream Commission
New Mexico Environment Department
New Mexico Institute of Mining and Technology
U.S. Bureau of Land Management
New Mexico Energy, Minerals and Natural Resources Department (NMEMNRD)
U.S. Bureau of Reclamation
U.S. Fish and Wildlife Service

SPONSORS

Patrick Gordon
New Mexico Geological Society Foundation
Charlie Nylander, WATERMATTERS, LLC
Rio Grande Restoration
Save Our Bosque Task Force
Sierra County Soil and Water Conservation District
Socorro County Soil and Water Conservation District
Representative Don Tripp

PRINCIPAL ORGANIZERS

Douglas Bland, Conference Chairman
Robert S. Bowman, Moderator and Technical Program Committee
Peggy S. Johnson, Technical Program Committee
Rolf Schmidt-Petersen, Technical Program Committee
Gina Dello Russo, Technical Program Committee
L. Greer Price, Guidebook Editor
Paul Bauer, Program Coordinator
Susan Welch, Logistics Coordinator
Barbara Fazio, Administrative Assistance

LOGISTICS

James Barker
Ruben Crespin
Lewis Gillard
Gretchen Hoffman
Mark Mansell
Geoff Rawling
Mike Timmons
Stacy Timmons
Dana Ulmer-Scholle
Patrick Walsh

PLANNING COMMITTEE

Consuelo Bokum, 1000 Friends of New Mexico
Robert S. Bowman, New Mexico Institute of Mining & Technology
Doug Boykin, State Forestry Division, NMEMNRD
Senator Carlos Cisneros
Gina Dello Russo, U.S. Fish & Wildlife Service
Gary Esslinger, Elephant Butte Irrigation District
Senator Dede Feldman
Ron Gardiner
Sterling Grogan, Middle Rio Grande Conservation District
Steve Harris, Rio Grande Restoration
Rolf M. Hechler, State Parks Division, NMEMNRD
Mike Inglis, Earth Data Analysis Center, University of New Mexico
Jim Jackson, New Mexico State Land Office
Janet Jarratt, New Mexico Water Assembly
John Matis, U.S. Bureau of Land Management
Doug McAda, U.S. Geological Survey
Gordon Meeks, Legislative Council Service, New Mexico State Legislature
Danny Milo, Office of Senator Jeff Bingaman
John J. Pfeil, Mining and Minerals Division, NMEMNRD
Connie Rupp, U.S. Bureau of Reclamation
Rolf Schmidt-Petersen, New Mexico Interstate Stream Commission
Peter A. Scholle, New Mexico Bureau of Geology and Mineral Resources
John Shomaker, John Shomaker & Associates, Inc.
Representative Mimi Stewart
Paul Tashjian, U.S. Fish & Wildlife Service
Representative Don Tripp
Erik Webb, Office of Senator Pete Domenici
Karl Wood, New Mexico Water Resources Research Institute

Contents

Preface... **vii**

An Introduction from the State Geologist—*Peter A. Scholle*... **1**

CHAPTER ONE
THE PHYSICAL AND HISTORICAL FRAMEWORK

Long River, Short Water: The Rio Grande Water Development Story –
Steve Harris... **7**
An Introduction to Rivers of the Arid Southwest –
Robert A. Mussetter... **14**
The Surface Water/Ground Water Connection –
Robert S. Bowman... **19**
Ecology of the Middle Rio Grande of New Mexico –
Mary J. Harner and Clifford N. Dahm... **23**
Managing Surface Waters on the Upper Rio Grande –
Rolf Schmidt-Petersen... **27**

CHAPTER TWO
THE MIDDLE RIO GRANDE TODAY

Infrastructure and Management of the Middle Rio Grande –
Leann Towne... **37**
The Middle Rio Grande Water Budget: A Debt Deferred –
Deborah L. Hathaway and Karen MacClune... **42**
Agriculture in the Middle Rio Grande Today –
Cecilia Rosacker-McCord and James McCord... **46**
The Middle Rio Grande Ecosystem through the San Acacia Reach –
Lisa M. Ellis... **50**
The Middle Rio Grande—Short on Water, Long on Legal Uncertainties –
G. Emlen Hall... **54**
Surface Water Management: Working within the Legal Framework –
Kevin G. Flanigan... **58**

CHAPTER THREE
CURRENT ISSUES ON THE MIDDLE RIO GRANDE

Drought and Middle Rio Grande Water Management Issues –
Ben Harding... **65**
A Killing "Cure"—Agricultural-to-Urban Water Transfers in the Middle
Rio Grande Basin – *Lisa Robert*... **69**
A Tool for Floodplain Management along the Rio Grande –
Matt Mitchell and Dick Kreiner... **73**
Salt Cedar Control: Exotic Species in the San Acacia Reach –
James Cleverly and Gina Dello Russo... **76**
The Endangered Species Act and the San Acacia Reach –
Jennifer M. Parody... **80**
Opportunities for Long-Term Bosque Preservation in the
San Acacia Reach–
Gina Dello Russo... **84**
How Science Can Provide Pathways to Solutions—
The Technical Toolbox –
Susan Kelly and Geoff Klise... **89**

CHAPTER FOUR
A VISION OF THE FUTURE

What We Stand to Lose in the San Acacia Floodway –
   Rolf Schmidt-Petersen and Peggy S. Johnson. . . **97**
Forging a Sustainable Water Policy in the Middle Rio Grande Valley—
a Downstream Perspective –
   Peggy S. Johnson and Mary Helen Follingstad. . . **102**
Water—Things to Do Now, and Do Better – *Frank Titus*. . . **107**
The Unintended Consequences of Water Conservation –
   Zohrab Samani and Rhonda Skaggs. . . **112**
Balancing the Budget: Options for the Middle Rio Grande's Future –
   Deborah L. Hathaway. . . **116**
Navigating the River of Our Future—The Rio Poco-Grande –
   William deBuys. . . **121**

List of Contributors. . . **131**

Photo Credits. . . **137**

Acronyms. . . **139**

Generalized Geologic Map of the Middle Rio Grande Region. . . **Inside back cover**

Preface

There is no subject of greater interest to New Mexicans than water: where we get it, how we use it, and how it is managed. Years of below-average precipitation have only increased that interest. Our focus for this volume is the Middle Rio Grande from San Acacia to Elephant Butte. This is a critical reach of the river. How water is managed in this region greatly affects water availability to the north and south, as well as our ability to meet legal obligations for water delivery. For decades we have fought and debated over the fate of water that, for some time now, has been fully appropriated. Three states and the Republic of Mexico have shared in that debate, and in the consequences of our decisions. For better or worse, we live today with a legacy of the immensely complex social and legal history of this river.

This volume is intended to provide a broad overview of these issues, including history, policy, legal framework, infrastructure, and management. The book is intended for a non-technical audience, providing information that is not otherwise easily found in a clear, understandable style and format. We hope it will serve as a reference for decision makers, policy makers, and the general public. Our authors were chosen for their ability to address these topics broadly and with authority, based on their expertise and experience.

This is the fifth volume we've produced in conjunction with our decision-makers field conferences. These conferences are designed to provide decision makers with an overview of earth science and related policy issues of interest and importance to all New Mexicans. We produced the first volume in 2001 on *Water, Watersheds, and Land Use in New Mexico*. This was followed by volumes on *New Mexico's Energy, Present and Future* in 2002, *Water Resources of the Lower Pecos Region* in 2003, and *Mining in New Mexico* in 2005. The conferences have been a resounding success, and the guidebooks we produce to accompany them have taken on a life and significance of their own. We've tried to focus on how science can aid in the decision-making process, and to provide a balanced view rather than a comprehensive one. If we've not provided answers to all of the questions, perhaps at least we have provoked significant thought and discussion.

We asked our authors to rely on fact rather than opinion, but the papers invariably reflect to some degree the views of their authors. Those views do not necessarily represent the voice of the New Mexico Bureau of Geology and Mineral Resources or our partner agencies. Our contributors are listed in the back of the volume, along with information about who they are and what they do. They, too, are an important resource and will remain involved in shaping the future of water in New Mexico for years to come.

Whatever that future may hold, it will require tough—and informed—decisions. It is our hope that this compilation will go far toward helping to inform those decisions. Our economic health, our environmental well-being, and the quality of life that we have come to take for granted in New Mexico all depend upon it.

—*The Editors*

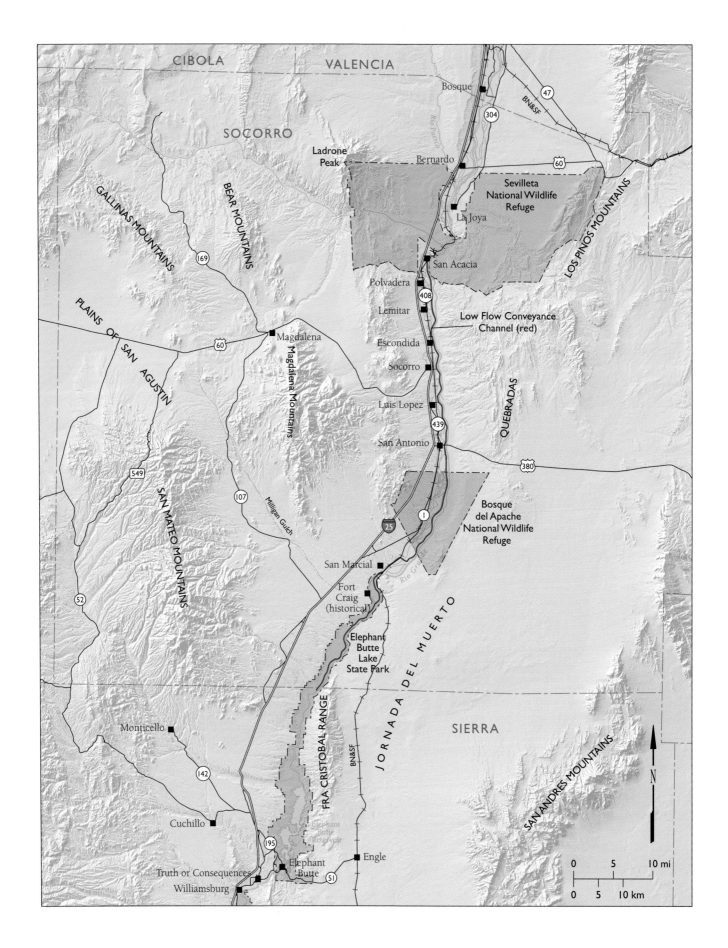

DECISION-MAKERS FIELD GUIDE 2007

An Introduction from the State Geologist

Peter A. Scholle, *New Mexico Bureau of Geology and Mineral Resources*

This year's decision-makers field conference, on water issues in the Middle Rio Grande region, again deals with some of the most difficult and contentious topics in New Mexico science and politics. Although this region has a relatively small population, it brings into play all the issues that face the water-supply situation throughout our state: It suffers the impacts of upstream use and abuse of water; it has habitat, endangered species, flood hazard, and water quality issues in its own reach; and it has to cope with the downstream obligations mandated by interstate compacts and international treaties. This volume is filled with papers that realistically discuss the constraints on the all-too-finite water supply in this arid state and this specific region. All the near-surface water in New Mexico is already owned by someone and is being used (or overused) for one productive purpose or another. We may be able to stretch uses through conservation and efficiency changes; we certainly can reallocate use from one sector to another (most typically from agriculture to urban or industrial uses), but we cannot generate new water. However, such major shifts in water uses, if undertaken, will have profound impacts on the character and landscapes of this state.

In addition to those rather grim current realities, we should remember two other factors. First, there is a growing scientific consensus that global warming is an undeniable reality. Although there are disagreements about how global climate change will be reflected in New Mexico, some models show increased aridity for this region and specifically less winter precipitation (and thus less storage of moisture in snowpack). In addition, several recent studies have shown that protracted periods of drought were much more common in this region over the past millennium than previously thought. So the predominantly wet conditions of the past century, under which our state has grown (and our interstate compacts were written), may be quite anomalous, adding to unsustainability of our water supplies. Paradoxically, as part of that pattern of change, many models predict more intense weather events, including heavy rainfall events, which could increase episodic flooding despite overall drought conditions. In other words, we may be heading for what we typically call "natural disasters."

I would like to expand on what is meant by a natural disaster. Hurricane Katrina was a recent major event that certainly earned this title. Yet like most such events, although Katrina was indeed a disaster, with the loss of more than 1,800 lives and $100-200 billion in insured and uninsured damages, it was only partially a "natural" disaster. The natural part was the hurricane; the unnatural part was the human and infrastructure damage that it caused, the lack of preparedness, and the simulated shock and awe in the political, regulatory, and emergency response communities at the aftermath.

Hurricanes have struck America's Gulf and Atlantic coastlines throughout recorded history (and well before that, as well). The strength and frequency of large hurricanes may be increasing, but the basic effects of hurricanes on coastlines have not changed appreciably over the centuries. In a natural setting, hurricanes erode shorelines, flood low-lying coastal areas, and flatten vegetation in their path. Most such damages are repaired in just a few years by sediment movement and vegetative regrowth. The disaster part of natural events comes from humans putting themselves and their infrastructure in harms way. Every geologist familiar with coastal processes could have predicted the inevitability of a hurricane disaster in New Orleans—not the year or the day, but the eventual inevitability. From a geological and engineering perspective, everything that could be done wrong was done wrong on the Mississippi River, and in and around New Orleans. The Mississippi River was forced to flow through New Orleans for more than a century after it would have shifted to the course occupied by the Atchafalaya River. Engineered levees along the Mississippi have not only kept the river in place, but also have prevented annual flooding and distribution of new sediment on the natural floodplains surrounding the river or along the coastline where it would have contributed to the formation of barrier islands. Instead, the sediment is now transported into deep water where it contributes not at all to coastal protection. Natural subsidence of the area has been compounded by subsurface water and oil withdrawals. Coastal dunes and shoreline mangrove forests, the natural defensive barriers for a coast, were compromised by extensive land clearing for housing developments. The broad wetlands and marshes that formed additional barriers along the coast were cut by networks of access channels that allowed entry, not just to drilling barges, but also to floodwaters. In short, the Mississippi delta region was the ultimate triumph of complex and incredibly expensive engineering projects over common sense and natural systems.

More than anything, government policy and public desires led to large populations and expensive infrastructure being placed in the inevitable path of storms. Subsidized federal flood insurance programs and tax incentives to spur economic growth, coupled with poor management practices, put people and structures into completely indefensible situations. Did anyone really believe that you could house hundreds of thousands of people in areas that lay 6 to 12 feet *below* sea level along a shoreline prone to hurricanes and not have a disaster some day? Did anyone really believe that walls built on mud, coupled with a series of easily floodable pump stations to remove water, were the solution to this problem? Does anyone now believe that rebuilding in the same sub-sea-level areas is any more sensible? Or that slightly higher walls will not, someday, be overtopped by slightly higher waters? Or that mandating that rebuilt houses be placed on 3 to 6-foot-high pads will solve flooding problems in still subsiding neighborhoods where people just recently drowned in their second-story attics?

Hurricane Katrina represents not only the costliest natural disaster in American history (by at least a factor of five over the next costliest event, Hurricane Andrew in 1992), it also represents what may be the nation's greatest communications failure as well. Every geoscientist who dealt with coastal processes, river systems, or even general sedimentology or oceanography understood the folly of the systems in place in New Orleans and could predict their ultimate fate. Many scientists at well-respected institutions, including the U.S. Geological Survey, most of the state geological surveys along the Gulf Coast, Louisiana State and other universities, conducted research and published scientific monographs on issues relevant to this disaster. But, for the most part, the concerns articulated by those scientists were not heard by the decision-making community. The technical jargon, the complexity of the issues, the unwillingness of most scientists to enter political frays, and the social dislocations and associated financial costs of the most rational solutions all blocked both communications and the acceptance of what was communicated. And it still does.

What does this all have to do with New Mexico and water-supply issues on the Rio Grande? I have purposely pointed my finger at a situation in which nobody here was involved to avoid the defensiveness that such accusations raise. But the parallels to what could happen in New Mexico in the future are compelling. The hydrologists, geologists, and other scientists working on climate and water supply in New Mexico are largely unified in their views that, in most areas of this state, we have reached or exceeded levels of water use that are sustainable even under current conditions. In addition, we face likely drought and climate change that will further compromise this supply. Nonetheless, in many areas we are allowing further development and are even providing substantial incentives for economic expansion. Furthermore, we are allowing a substantial part of that development to occur on the floodplains of our major rivers. How does this differ from encouraging building in coastal hazard zones or, to exaggerate ever so slightly, to zoning the top of Mount St. Helens for residential development? The natural hazards of drought and flooding in New Mexico will probably not lead to the massive deaths of Hurricane Katrina, but they could lead to huge economic losses, at personal, corporate, and governmental levels, comparable to the Katrina event. And they could lead to social disruptions and a population exodus that also compares to Katrina and New Orleans.

The use of floodplains, in particular, presents a series of problems that may put us on the very same slippery slope of engineered projects that led to the Katrina disaster. Floodplains, as the name implies, are natural overflow areas during those times when a river has to carry more water than can be accommodated by its channel. Floodplains are excellent sites for agriculture and for storage of things that can be quickly and easily moved. Once the decision is made to allow other development on floodplains, including residential housing, industry, or commerce, there are immobile infrastructure, personal property, and lives at risk. The value of those lives and properties is used to justify the construction of very expensive protective engineering structure (dams, levees, concrete-lined channels, and the like), and that puts us on a permanent path of working against nature rather than with it.

Engineered structures are designed to meet specific conditions, but conditions change with time. Construction of upstream levees funnels more water through channels and adds stress on levees in downstream areas. Urbanization upstream (read covering the land with concrete and asphalt) reduces water infiltration during rainstorms. The water is instead funneled to the river through storm drains and concrete channels. This increases the flashiness of flash floods and again puts new stresses on levee systems. Climate change also contributes to the problem. What once were 500-year events become 100-year or even 50-year events, and thus higher levees need to be built and maintained, or people's lives and property will be in ever greater danger from flooding. The cost of such engineering is great, the structures require perpetual maintenance and upgrading, and such projects are always fraught with danger of failure during unanticipated large events. But it should be pointed out again that natural disasters are true disasters only because we allowed inappropriate land uses in the first place.

Of course, it is a legitimate role for government to provide opportunities and incentives for economic and social development, but such incentives should be applied wisely and with a view to long-term rather than short-term gains. Viewed from a long-term perspective, projects that fight nature, whether they involve coastal beach replenishment in the face of rising sea levels or building river levees in the face of increased flooding, are unsustainable. Government should have the wisdom to create solutions that make long-term sense, both economically and in terms of public safety. That requires intelligent consideration of water and zoning issues and the full incorporation of scientific and technical knowledge into such decision-making.

So my request to you as decision makers and as citizens is:

- Please read the papers in this volume, grasp the consensus on the issues discussed, and think of those issues as you help shape policy for New Mexico.

- Continue the dialog between scientists and policy makers that we hope this conference will start (and please remember: Scientists are those shy, quiet people who stood along the wall during your senior prom, so they may need a little coaxing).

- Encourage and financially support the gathering of the fundamental scientific data needed to fully understand water supply in this state as well as to create rational water plans. We have learned much in past decades, but this is a large state with much still to be learned.

- To the maximum degree possible, support natural, not engineered, systems (see the last paper in this volume, by Bill deBuys, for further reasons to do this). Engineered systems limit our flexibility and the ability of natural systems to function naturally.

- Think of water as you make decisions not just about water, but as you deal with growth and development as well. Think of water as the limiting factor on everything that happens in this Land of Enchantment, and you will help to keep New Mexico exactly that.

CHAPTER ONE

THE PHYSICAL AND HISTORICAL FRAMEWORK

DECISION-MAKERS
FIELD CONFERENCE 2007
San Acacia to Elephant Butte

CHAPTER ONE

The Rio Grande near San Acacia

Long River, Short Water: The Rio Grande Water Development Story

Steve Harris, *Rio Grande Restoration*

The rich alluvial valleys of the Rio Grande have supported agriculture for nearly a millennium. In a semiarid land, through capricious swings of drought and flood, the soils and water of the river have nurtured substantial civilizations and inspired cultural traditions that continue to enrich modern New Mexico.

The river's first farmers were Pueblo people. One scholar estimates sixteenth century Pueblo populations as great as 80,000 persons, in 100 villages, making the Rio Grande Pueblo civilization the greatest concentration of settled farming villages in the American Southwest. The first farmers were irrigators, though they appear to have relied more upon such elegant moisture-conserving techniques as water-retaining terraces, cobble mulches, and self-contained "waffle gardens" than on intensive dam and canal systems. Because labor requirements were high, Puebloan agriculture was a necessarily cooperative venture. Their use of resources was likely governed less by political control than by traditional sacred relationships to land, sky, and river.

Though at times as many as 30,000 acres may have been cultivated, Pueblo impacts to the stream would seem modest to modern-day farmers. European explorers marveled at the quality and abundant yields of Pueblo plantings. Indeed, the earliest Spanish colonists might have perished without the surpluses of corn and beans laid up by the first farmers.

THE SPANISH ENTRADA

In 1591 the frontier of European expansion reached the Rio Grande. Driven hither by the quest for wealth and Christian evangelism, Spanish conquistadors found both minerals and souls were hard to come by. Still, compared with the expanse of desert to the south, the *Rio Bravo del Norte* offered rich soils and abundant water. Encouraged by grants of land from the royal government, a stream of Spanish immigrants, mainly impoverished exiles, flowed into the region over a 250-year period and conquered the north.

An advanced irrigation technology came with them, in the form of acequia agriculture. Headings of rock and brush and hand dug canals served to turn water onto pastures and cultivated fields. In the few large towns royal governments, *alcaldes* and *ayuntamientos*, governed colonial affairs, including the division of water, whereas in the many small villages an indigenous water democracy maintained a cooperative governance by acequia *majordomos* and *comisionados*.

Spanish traditions, grafted onto new world realities, suggest that water users shared the benefits and losses of the variable supply. Priority of first use was respected, though not with such exclusivity as in the modern appropriation doctrine. When water was scarce, demonstrated needs (especially for drinking water and stock watering) and concepts of fairness were often the basis for an allocation decision. These traditions continue to carry legal weight, as formalized in the 1907 Territorial Water Code.

By 1821 acequia agriculture, both Pueblo and Hispanic, had grown until it involved no more than 150,000 scattered acres between Taos and Tome. Possessed of pragmatic technologies for water control, acequia irrigators relied on cooperation, hard labor, and the will of God to bestow the blessings of the river to their land.

ANGLO-AMERICAN CONQUEST

The United States' conquest of Mexico's northern territories in 1848 signaled a profound transformation of the localized, cooperative traditions of water development and governance on the Rio Grande. In their place, industrial technologies and the U.S. doctrine of "Manifest Destiny" established a model for national possession of western lands. Henceforth, a restless hoard of speculators swept over North America's vast western empire to turn minerals, timber, grass, and water into dollars.

Agriculture was a key part of the U.S. national policy of rapid immigration and development of the West's natural resources. However, to farm successfully beyond the one hundredth meridian required irrigation. From a few successful experiments with large-scale irrigated farming, ambitious water diversion projects spread like prairie fire to every river valley in the West. Quickly, possession of western rivers was granted not to the owners of the land or the communities through which they flowed, but to the persons who built the works that diverted them.

The growth of irrigation from the Rio Grande typifies

the explosiveness of this process. In 1850 Rio Grande farms from San Luis Valley of Colorado to El Paso Valley, Texas, totaled less than 200,000 acres. By the time the temporary Rio Grande Compact was signed in 1929, irrigation in the basin encompassed more than 1,000,000 acres.

COLORADO DEVELOPMENT

The vast, fertile, high-elevation San Luis Valley was not settled until 1851, when Hispano settlers spilled northward from the Taos region and were soon joined in 1878 by westering Mormon farmers. The opening of the Denver and Rio Grande Railroad to Alamosa in 1878 ignited the biggest of the Rio Grande's "big barbecues." Between 1880 and 1890 British speculators financed five large canals dug by mule-drawn scrapers. Cumulatively these canals could divert almost 5,000 cubic feet per second, virtually the entire spring runoff of the main river. Colorado attempted to secure legally the natural advantage of its location at the top of the hydraulic system, claiming an unimpeded right

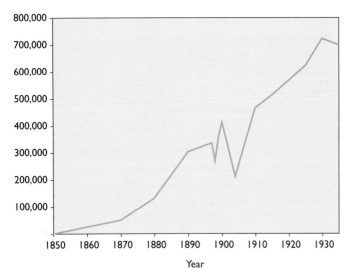

Irrigated acreage in San Luis Valley, Colorado, illustrating the growth in upstream water diversions, which has dramatically impacted New Mexico water supplies from the Rio Grande.

to the waters that arose within the state. Though disabused of their "doctrine of sovereignty" by the Supreme Court in 1917, the great deeds of irrigation development were largely already done. A general stream adjudication completed in 1891 showed that more than 300,000 acres had been placed under ditch. The impact of this rapid and extensive development would be immediately felt by downstream water users and would impact the course of Rio Grande history for more than a century to come.

NEW MEXICO

Meanwhile and farther south, expansion of irrigation was also occurring, though at a more restrained pace. In the Rio Arriba (that stretch of the river north of La

"The Big Barbecue"—a term applied by historian Charles Wilkinson to the rapid conversion of Western resources into wealth in the late nineteenth century. This log jam is on Embudo Creek, circa 1915. The timber was headed to the Rio Grande, through White Rock Canyon, and ultimately by rail to Albuquerque.

Bajada), irrigation had long since reached its full development, with perhaps 100,000 acres applying water from the several hundred acequias originating in tributary streams. In the Rio Abajo (middle Rio Grande), after the settling of Tome in 1739, Spanish–Mexican farming grew continuously until 1880, when it comprised about 124,000 acres, irrigated by more than 70 traditional acequias.

When the Atchison, Topeka, and Santa Fe Railroad reached Santa Fe in 1880, New Mexico also connected to commerce with the wider nation. Industrial-scale grazing began to make economic sense, and the north experienced a sheep-grazing boom. Likewise, thousands of acres of Sangre de Cristo forests were harvested and boomed down the Rio Grande to Embudo Station for use as railroad ties. A result of this large-scale timber development, abetted by a national policy of fire suppression, was that the Rio Grande's high-elevation watersheds, upon which the region's acequias depended, were rapidly transformed. Snowmelt in the Rio Grande's tributaries began to come more quickly and in reduced volume, its hydrographs less attenuated into the summer season. Sediments released by logged-off forests and grazed-off grasslands aggraded river channels, which, with reduced peak flows from Colorado diversion, were less able to maintain themselves.

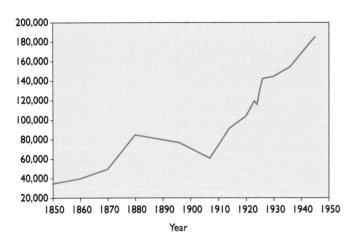

Irrigated acreage in Mesilla, Rincon, El Paso, and Juarez valleys, New Mexico, Texas, and Mexico, illustrating the sharp growth in irrigated acreage after construction of Rio Grande Project canals below Elephant Butte.

THE RIO GRANDE PROJECT

Water development in the Mesilla and El Paso/Juarez Valleys followed the pattern of the Middle valley, up to a point. An 1858 survey portrayed acequia farming in the region utilizing about 10,000 acres. By 1880 as many as 25,000 acres may have been irrigated, and speculators had their eyes on more.

But in 1890 what would become a nine-year drought descended on the Rio Grande. The years of low snowpack dampened the onrush of development and led to substantial, if temporary, declines in irrigated acreage. The San Luis Valley took whatever water it could, and the middle valley often diverted what remained of the river. In the El Paso/Juarez Valley, these upstream diversions compounded the drought and caused the region's famous vineyards to wither and die. About the same time, and far downstream, steamboat navigation of the Rio Grande below Laredo ceased forever.

In 1889 the Rio Grande Dam and Irrigation Company of Mesilla, New Mexico, was incorporated and proposed the construction of a reservoir and canal system to irrigate 530,000 acres in Mesilla Valley. By 1895 the company had received approval for a reservoir right of way from the Secretary of Interior. El Paso/Juarez farmers responded to the impending, profound water shortage with outrage, leading Mexico to file damage claims for $35,000,000 against the United States. Here was a serious diplomatic breach, and the International Boundary Commission was assigned to study the problem. The problem, their report concluded, was that the border area "suffered from the increased use of water in Colorado."

After a decade of diplomatic wrangling border officials in the two countries determined that the solution was to build a storage dam at the El Paso Narrows. Suddenly there were two conflicting reservoir proposals on the table. El Paso/Juarez interests were utterly opposed to the Elephant Butte Project; it was too far away, and the speculators were geographically too well-positioned to intercept and control the water. New Mexico strenuously opposed the El Paso dam, which provided no water storage for proposed developments around Mesilla.

By 1906 an unlikely, but momentous, series of events had occurred, resulting in resolution of the 18-year-old problem:

- First, a territorial court rejected U.S./Mexico arguments that the private dam would illegally interfere with navigation of the river. Resolving the litigation in the case's third review, the U.S. Supreme Court ruled that the Rio Grande Dam and Irrigation Company had waited too long to begin construction on the Elephant Butte Project. Its patent was thereby repealed, clearing the way for a single federal project.

- The International Treaty of 1906 "to equitably distribute the waters of the Rio Grande" was signed by both nations, who assented to a three-way split of the lower Rio Grande. In the 1906 treaty, Mexico settled for a 60,000 acre-feet guarantee, delivered from the reservoir each year "except in times of extraordinary drought."

- After more than twenty years of stormy debate on how best to advance western irrigation, Congress passed the landmark Reclamation Act of 1902, creating a firm policy of federal financing (and control) of irrigation development. In exchange for the territorial engineer's granting 730,000 acre-feet of water rights to the federal Rio Grande Project, New Mexico induced its border neighbors to accept the Elephant Butte Reservoir site.

- In 1908 "all the unappropriated waters of the Rio

Grande and its tributaries" were decreed to the Reclamation Service. In 1911 construction began on the project; by 1916 a completed Elephant Butte Reservoir, one of the Reclamation Service's first projects, began to store the 2.6 million acre-feet for which it was designed.

RIO GRANDE COMPACTS

A federal embargo, declared by the Secretary of Interior in 1896, prevented Colorado from constructing the dam it desired at Wagon Wheel Gap but otherwise did little to reduce the development of new irrigation. While Supreme Court cases, international treaties, and major reservoirs were being negotiated, the sovereign state of Colorado continued expanding its exploitation of the Rio Grande. In 1924 San Luis Valley water commissioners reported a total of 621,826 acres under irrigation, up from 213,210 in 1896.

Fervid ambition was the only factor governing either the Rio Grande Project or San Luis Valley development, as by 1929 irrigation had increased dramatically, both downstream of Elephant Butte and in Colorado. Even with a period of abundant snows in the 1920s, the Rio Grande regularly disappeared at late season in the middle valley. It was becoming clear that only an interstate agreement could bank the fires of the Rio Grande's big barbecue.

In 1925 the embargo was lifted, as Colorado, New Mexico, and Texas agreed to seek an interstate compact equitably dividing the Rio Grande among them. A temporary compact was put into effect in 1929 to freeze the apportionment at then-current levels. To effect a permanent agreement, and to determine the nature of the water supply and the relationship of each segment's demands, required the collection and analysis of water-use data. Under auspices of the National Resources Committee, national and state scientists conducted an exhaustive joint investigation to determine the facts needed to equitably balance each section's inflow, outflow, and demand.

Completed in 1935, the National Resources Committee's regional planning report provided the foundation for a definitive negotiation among the states. The report acknowledged that the Rio Grande was at or beyond the limits of the water it could be expected to provide: "...with the available water resources of the Rio Grande apparently fully appropriated, the approval of any new projects involving additional drafts upon those resources seem to point inevitably to further conflict...."

The three sections' bottom line for negotiations to resolve the conflict was clear. Colorado would consolidate the dramatic gains it had made and perhaps be allowed to build a storage reservoir. The middle valley, too, would need its own reservoir to regulate late season supplies. Rio Grande Project users wanted assurance that the others would leave enough water to supply their needs and aspirations. Thus informed, the compact commissioners and their legal and technical advisers negotiated, over three years, a set of delivery schedules and various caveats to fix their irrigation demands to the fluctuating supply, resulting in the present Rio Grande Compact.

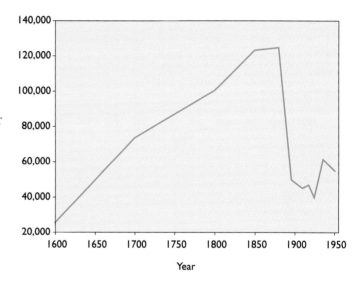

Irrigated acreage in Middle Rio Grande, Cochiti to San Marcial, New Mexico, illustrating precipitous declines in irrigation when sediment burdens and aggrading channels combined to waterlog more than half of the historically irrigated lands. Note also the partial reclamation resulting from MRGCD and federal projects.

MIDDLE RIO GRANDE PROJECTS

With or without interstate accords, the Middle Rio Grande valley increasingly found itself in a desperate position: bracketed by two thirsty, fast-moving competitors, one of which had recently vouchsafed a claim to virtually the entire flow of the river. Its organizing principle, the acequia system, isolated it from the power politics of large-scale irrigation. It had no reservoir to regulate a diminishing river. Its economy was also declining in lockstep with intensifying competition from the other two regions and the deteriorating condition of its lands.

Not only were its supplies of river water diminishing, but in the mid-river, the Rio Grande was leaking into the fields. By 1896 irrigated agriculture had declined from a high of 125,000 acres to 50,000 acres. The deadly combination of silt from deteriorating watersheds being deposited in the channel, reduced channel-forming flows from water intercepted by upstream irrigators, and its own flood irrigation practices clogging the Rio Grande, raised its channel above the elevation of the surrounding floodplain and seeped into much previously productive land. A 1918 state engineer inventory of middle valley conditions revealed nearly 60,000 acres of waterlogged and alkali-salted former farmland. In addition, the aggrading river flooded with increasing frequency, playing havoc with earthen irrigation works and cutting sandy channels across the beleaguered farms.

Following several abortive local efforts to finance a drainage system, a joint Bureau of Reclamation–state engineer commission proposed a solution: a comprehensive plan for drainage, flood control, and channel rectification, complete with a (180,000 acre-foot) storage dam, and a consolidated series of diversion dams and main canals to replace the primitive diversions and ditches. Because such a project promised to be extremely costly, farmers hoped that it might be financed through the federal Reclamation Fund. However, because New Mexico had already received a substantial share of such funds in the lower river, another mechanism would have to be found.

An intensive lobbying effort by Albuquerque and rural leaders convinced the state legislature to approve the Conservancy District Act of 1923. Districts created pursuant to this act were to be organized and administered by a state district court, upon petition by 100 landowners. After two petitions to the district court, the Middle Rio Grande Conservancy District was successfully organized in 1925. The Middle Rio Grande Conservancy District was to serve all lands in the floodplain of the Rio Grande between Cochiti and San Marcial and thus could add its assessments for flood protection to the property taxes collected from residents of Sandoval, Bernalillo, Valencia, and Socorro Counties. The 130,000-plus acres projected to receive irrigation water from the district would be levied additional assessments to construct and maintain those works. Included in the conservancy district were 28,500-plus acres of Pueblo Indian lands, for which Congress appropriated more than $1.5 million to cover construction costs on Indian lands.

At the outset, some of the district's intended beneficiaries opposed creation of the district, and many remained suspicious of its subsequent arrangements. The Middle Rio Grande Conservancy District was a new and powerful political subdivision of the state, with extensive powers to make regulations, levy taxes, condemn and own lands and water rights, salvage water, remove or relocate structures, fill lands, retard silt, re-engineer stream channels, construct drains, dams, levees, canals, roads, bridges, stream gages, and electric power plants. Its water rights were to be exempt from forfeiture under state law or taking by other political subdivisions. As it condemned existing acequias, the conservancy was required to supply its parciantes (shareholders) with the water entitlements to which they had become accustomed.

In 1928 the Middle Rio Grande Conservancy District submitted its "Official Plan for Flood Control, Drainage and Irrigation" to the district court. Construction soon began on four diversion dams,

The same bend in the Rio Grande near San Acacia in 1905 (left), showing a broad channel, flood-swept sandbar, and large wetland in distance, and again in 1989 (right). By 1989 the river flowed only partly in its native channel, with a levee, riverside drain, and main canal joining the Santa Fe railroad along its neatly engineered course. A thicket of invasive salt cedar now covers the floodplain and confines the river channel, which has narrowed by 200 feet and aggraded 15 feet in the 84-year interval.

their connecting main canals, a valley-wide system of riverside and interior drains, and El Vado Reservoir. Initial tax assessments appeared to be substantial enough to satisfy court-appointed appraisers that the district could service the bonds it let to finance the estimated $12 million cost of construction.

Unintended consequences from the conservancy district's project were substantial. Water supply to some ditches was interrupted during construction, reducing their parciantes' ability to farm for one or two years. Several thousand acres of farmland were condemned for rights of way to the drains and canals. A number of irrigators failed to make their annual assessment payments, resulting in foreclosure of some 34,000 acres by the state. Other ratepayers felt that the original glowing promises of project benefits had been overstated. Certainly, the flood control works did not prevent the devastation of the Socorro division by the 1937 and 1941 floods. Siltation and aggradation of the channel continued to plague the river. Additionally, there were, and continue to be, assertions that the broad powers of the Middle Rio Grande Conservancy District inhibited the state engineer's authority to administer individual water rights priorities.

At least some of the project's benefits were realized: as many as 20,000 acres were drained and (at least temporarily) reclaimed for farming. The construction of El Vado Reservoir succeeded in extending late-season water supply to the district's four irrigation sections.

Further bedeviling the Middle Rio Grande valley, an interregional conflict erupted when the district began to fill El Vado Reservoir for the first time in 1935. Texas sued New Mexico and the Middle Rio Grande Conservancy District in the U.S. Supreme Court, claiming that the defendants were impairing the water supply in Elephant Butte, in violation of the 1929 compact. This litigation was dismissed after the 1938 compact was signed by the three states. By 1941 Congress was authorizing the Corps of Engineers to study the still-unmet drainage and flood control needs of the region.

Nor did the district quite succeed in rescuing its farmers from financial woes. In danger of defaulting on the bonds that financed the project, the district appealed to Congress in 1948 for relief of its debts and the rehabilitation and further improvement of its dam, diversions, and levees. The Flood Control Act of 1948 authorized a Middle Rio Grande Project through the U.S. Army Corps of Engineers and appropriated $15 million through the Bureau of Reclamation to provide the district with debt relief and another round of middle valley "improvements." These improvements included 300,000 jetty jacks to straighten and confine the river channel. Reclamation was to hold the titles to the water rights and capital works as security for repayment of this crucial federal investment.

The focus of the Middle Rio Grande Project was primarily on the worsening siltation problems. It gave the Bureau of Reclamation the authority to maintain an open river and funds to channelize 127 miles of river from Velarde to Elephant Butte, resulting in the placement of over 300,000 jetty jacks to confine the river channel over the next 20 years. The Flood Control Act of 1950 authorized more than $50 million to the Corps of Engineers to construct flood and sediment control reservoirs, the crux of a strategy to radically reduce sediment inputs to the valley. Abiquiu Reservoir on the Rio Chama and Jemez Canyon Dam were completed for this purpose in 1954. Construction of the additionally contemplated sediment and flood control reservoirs was deferred until a Rio Grande Reservoir Regulation Plan was negotiated to the satisfaction of Colorado and Texas. In 1965 the corps began work on its own Middle Rio Grande Project, completing Galisteo Dam in 1970 and Cochiti Dam in 1975.

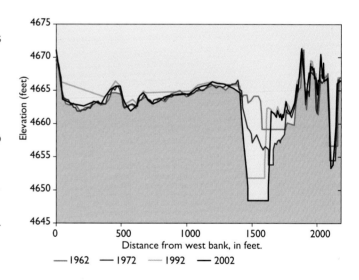

Time series of Rio Grande channel forms near San Acacia, showing a dramatic narrowing and incising of the river channel.

Cochiti has been a particularly significant development in the history of the Rio Grande. Its location on Cochiti Pueblo lands displaced much of the floodplain farming that was culturally and economically critical to the tribe. During its construction, a sacred site important to Cochiti and its neighboring Pueblos was carelessly destroyed. Then, when the reservoir filled, seepage below the dam waterlogged what remained of the Pueblo's farmlands. One intended benefit, sediment abatement in the middle valley, resolved itself poorly as sediment-starved waters began to progressively scour the river channel downstream. Cochiti Reservoir's position athwart the mouth of White Rock Canyon isolated the river below from a natural refugium for aquatic species, contributing to the decline and endangered status of the Rio Grande silvery minnow. It has, however, kept its flood control promise, intercepting potentially damaging floods in 1979, 1985, and 1995.

RECENT TIMES

The Rio Grande Compact, with its cornerstones of sound science and frank if difficult negotiations, has served to moderate the consequences of the Rio Grande's century-and-a-half-long development orgy. The compact, first administered in 1940, is the foundation of today's "law of the river," which also includes a welter of contracts between special water districts and the Bureau of Reclamation, water rights administration in three states, and the decisions made by thousands of individual water users and their districts.

Unfortunately, both Colorado and New Mexico have found that they cannot always reliably comply with the compact's downstream delivery requirements. And so, when threats of harm cannot otherwise be reconciled, the courts are standing by.

During the severe drought of the 1950s the New Mexico State Engineer and the Middle Rio Grande Conservancy District again found themselves before the U.S. Supreme Court to answer for a water debit that had exceeded its 200,000 acre-foot limit allowed by the compact. The proximate cause of this action was a renewed assertion that New Mexico was storing water in El Vado Reservoir in violation of the Rio Grande Compact. The U.S. Supreme Court dismissed this litigation, ruling that Texas had failed to name an indispensable party, the United States in its capacity as trustee for the six Middle Rio Grande Pueblos and their water rights.

By 1956 New Mexico's debit had grown to more than 500,000 acre-feet, before rehabilitation of waterlogged lands undertaken by the Middle Rio Grande Project could produce additional water flows. The Low Flow Conveyance Channel, which began operating in 1953 in an attempt to make New Mexico's compact deliveries more reliable, provided a bit of the long sought drainage objective. The channel did, in fact, produce water and reduced the state's compact debit, but today it serves mainly to drain a Rio Grande channel perched dozens of feet above the surrounding floodplain.

The San Juan–Chama Project also helped New Mexico to become compact compliant. The project began diverting water from the San Juan River basin into Heron Reservoir in 1972, and started releases in 1974. One of its original stated purposes was to provide 42,500 acre-feet per year to "replace previous and anticipated [Rio Grande] basin depletions caused by miscellaneous uses." San Juan–Chama Project water has helped ease the state's chronic non-compliance, sending an average of 30,000 annual acre-feet downstream, effectively offsetting water sucked from the river by ground water pumping.

During the same period, Colorado's accrued debit swelled to almost 1,000,000 acre-feet. In 1966 New Mexico joined Texas in a Supreme Court suit that resulted in an agreement by Colorado to begin to reduce its huge deficit. Across-the-board curtailment of San Luis Valley irrigation forced farmers to conserve and reduce their water applications.

The federal Closed Basin Project, which began salvaging shallow ground water in 1984, was designed to reduce the state's accrued debit. Nevertheless, it was not until Elephant Butte Reservoir spilled in a very wet 1985 that Colorado's 30-year-old water debt was forgiven.

Over the past 20 years, with a blessing of abundant snowpacks, both Colorado and New Mexico have maintained compliance with the water delivery requirements of the Rio Grande Compact.

FINALE

Stretched thin by the dizzying pace and magnitude of water development, the Rio Grande/Rio Bravo basin remains enmeshed in a perpetual conundrum: there are simply more claims to Rio Grande water than the river can reasonably be expected to provide. Its core ecology, the very structure, and function of the river have been profoundly altered, with unfortunate outcomes: It ceases to flow at the behest of unrestrained economics; its leveed banks armor continuously narrowing and aggrading channels, disrupting the conveyance of water; biodiversity continues to decline. Successive engineering projects have disrupted the productivity of its adjoining lands, now beset by invading plants and, consequently, fire.

Another chapter of the saga will be written by the present generation. Access to the river by growing, thirsty urban populations and the emergence of concern for the fate of Rio Grande ecosystems have joined the perennial contenders for the limited supply. The planned use of San Juan–Chama Project water by Albuquerque, Santa Fe, and other communities suggests that a challenging new version of the intricate old balancing act lies just ahead.

An Introduction to Rivers of the Arid Southwest

Robert A. Mussetter, *Mussetter Engineering, Inc.*

One of the most startling paradoxes of the world's drylands is that, although they are lands of little rain, the details of their surfaces are mostly the products of the action of rivers. To understand the natural environments of drylands, deserts, arid, and semiarid regions of the earth is to understand the processes and forms of their rivers.

—William L. Graf, *Fluvial Processes in Dryland Rivers*

Rivers are the primary conduits by which water and the products of continental weathering are delivered to the oceans. Flowing water provides the energy that moves the weathered soil and rock material, along with organic matter that is important to the ecosystem, downstream through the river system. The water is derived from precipitation that falls on the earth's surface and continues through the hydrologic cycle in a variety of forms, including surface runoff, infiltration into the ground water system, and evaporation and transpiration back into the atmosphere where it contributes to later precipitation. At many locations, much of the water that appears as river flow is derived from the ground water system, which provides storage that sustains stream flow during dry periods. In rivers with mountainous headwaters, winter precipitation is temporarily stored as snow until the seasonal temperatures rise above the melting point, releasing it into the local stream system.

The relative contribution of each of the above factors to the flow at any particular location and time varies considerably. Intense local storms can cause extreme changes in stream flow over very short time periods on the order of minutes to hours, whereas seasonal changes in temperature and precipitation result in predictable patterns in stream flow over the course of the year. Longer-term climate variability causes both cyclical variability and random fluctuations in the amount of stream flow over a few to many years. Global warming may also affect long-term stream flow patterns.

In addition to stream flow that is mostly climate driven, the physical characteristics of rivers are strongly influenced by the basin geology that controls the character of the sediment in the river bed and banks, and provides structural controls on the alignment, slope, shape, and dynamic behavior of the river. From the human perspective, these characteristics are critical because they affect the capacity of the river to safely carry flood flows and its ability to continue carrying lower flows when the water is essential to ecosystem health and water supply needs. The dynamic (erosional and depositional) behavior of the river is also important because it affects the safety of infrastructure, particularly during high-flow periods.

Because water is key to nearly all physical and biological processes, river corridors are a critical component of the natural ecosystem. This is nowhere more true than in arid regions, where the health of the riparian corridor is tied directly to the amount and timing of flow that mostly originates from outside the local area. Natural stream flow patterns in arid rivers tend to be more variable than in temperate- or humid-zone rivers, and the species that are found in arid rivers generally have adapted to this high variability. Human use of the water often alters the flow patterns and thus the health of the ecosystem. Understanding these effects and managing water use to support human needs and protect public safety during flooding, while minimizing ecosystem impacts, are key challenges for river managers.

HYDROLOGY OF THE RIO GRANDE

All of the above factors that affect stream flow are important in the Rio Grande basin. The Rio Grande, originating in the mountains of southern Colorado, is the fifth longest river in North America, with a total contributing drainage area of about 176,000 square miles, including the arid lands of New Mexico, Texas, and northern Mexico. Annual precipitation in part of the upper basin averages nearly 40 inches per year, whereas much of the middle and lower basin receives less than 10 inches per year. Typical of arid rivers, the Rio Grande tends to lose flow in the downstream direction, and did so even under native (pre-European settlement) conditions, due to infiltration and evapotranspiration. Upstream water use and to a lesser degree increased evapotranspiration losses caused by the proliferation of salt cedar and other non-native vegetation now cause river flows to decrease in the downstream direction even more so than under native conditions.

THE PHYSICAL AND HISTORICAL FRAMEWORK

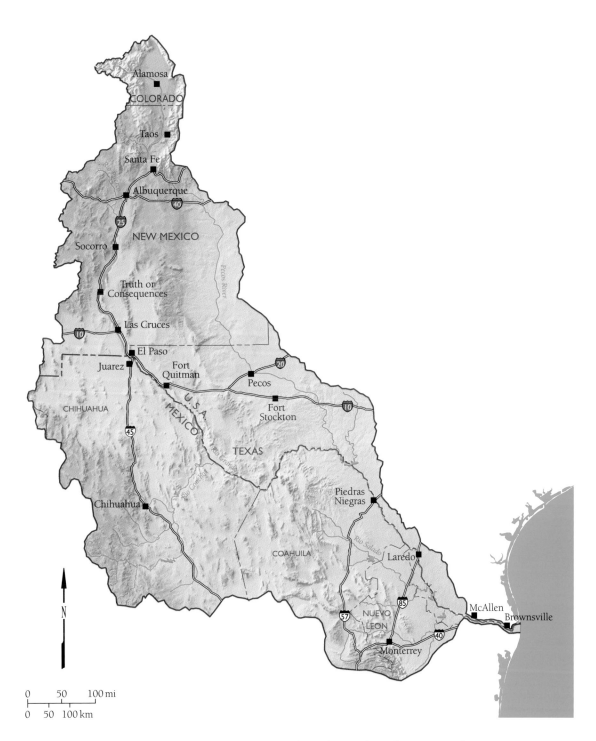

The Rio Grande watershed. The Rio Grande drains areas of three states and parts of northern Mexico. Below El Paso the river forms the international boundary between the U.S. and Mexico.

The annual water volume of the Rio Grande that flows past the Otowi stream gage above Cochiti Reservoir averaged about 1.1 million acre-feet under native conditions, and the current long-term average is about the same. The volume varies significantly from year to year. The lowest recorded volume at Otowi since 1895 was about 360,000 acre-feet (1904), about one-third of average, and the highest recorded volume was about 2.4 million acre-feet (1940), about 220 percent of average. Over the more recent six-year period (water years 2000 through 2005), the average volume was about 760,000 acre-feet, about 30 percent

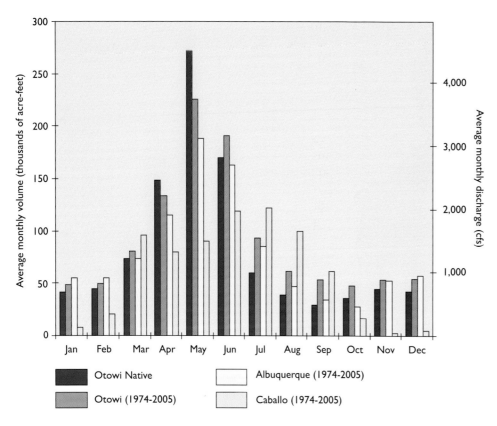

Average monthly flows on the Rio Grande, at four locations, as noted.

less than the long-term average.

Native flow volume estimates unfortunately are not available for the middle and lower reaches. Since 1973 when Cochiti Dam was completed, the amount of water passing Albuquerque averaged slightly less than 1 million acre-feet per year, and this decreases to about 0.73 million acre-feet below Caballo Dam. The post-Cochiti average at Albuquerque was about 40 percent greater than the 30-year average from the early 1940s to early 1970s. The difference between the two periods is due to both direct human influence (e.g., imported flows from the San Juan–Chama Project, completed in 1971), and natural causes (i.e., extended periods of below average precipitation in the 1940s through mid-1970s and above average precipitation in the 1980s through mid-1990s.)

Under both native and current conditions, the bulk of the annual flow volume is derived from snowmelt from the upper basin; thus, the highest sustained flows typically occur during late spring and early summer. Prior to significant water development, about 60 percent of the annual native volume in the upper reaches occurred during April through June, and less than 15 percent occurred during July through September. Seasonal patterns in the downstream reaches were probably similar to the upstream reaches. Human-induced factors, including the water storage system, imported flows, urbanization and channelization, and increased evapotranspiration as a result of non-native vegetation, now affect the seasonal flow patterns, with the magnitude of the effects increasing in the downstream direction. In the upper and middle reaches, about half of the annual flow volume now occurs during April through June, and the July through September percentages have increased to 15–20 percent. In contrast, only about 40 percent of the annual flow below Caballo Dam now occurs during May through June, but the amount during July through September represents about 40 percent of the annual volume. The reasons for this difference are complex, but they are mostly driven by upstream, man-induced changes. Late summer, monsoon season runoff from the middle and lower basin is also a factor.

GEOMORPHOLOGY OF THE RIO GRANDE

The geomorphic characteristics of rivers represent the integrated effects of the physical factors in the basin and drainage network that include the quantity and timing of flow (i.e., the hydrology), drainage basin geology, riparian vegetation, and human modifications. Rivers tend to adjust their gradient, plan form (their down-valley and cross-valley alignment), cross sectional shape, and the sediment size toward a state of dynamic equilibrium (or long-term balance) with the upstream water and sediment supply. In this regard, it is a common premise that alluvial rivers tend to adjust their cross sectional size so that the capacity at which water will spill onto the floodplain (often referred to as the bankfull capacity) corresponds to the maximum discharge that occurs for a few to several days each year. Because of the highly sporadic nature of the runoff, however, arid rivers such as the Rio Grande may be close to equilibrium for only a very small percentage of the time; thus, the in-channel capacity can vary considerably from the annual peak flows.

High flows that occur on an annual or less frequent basis are the most important in forming and maintaining the river channel. Erosion, transport, and deposition of sediment during these high flows are, in turn, key to maintaining certain aspects of the ecosystem. In river systems like the Rio Grande, high flows result from both snowmelt runoff and runoff from intense rainstorms in the lower basin. Although the storm-driven flows can be as high, or higher than, the snowmelt-driven flows, the latter tend to occur for much longer durations; thus, they do more work in shaping and modifying the channel. Before construction of Cochiti Dam, the annual maximum snowmelt-driven discharge near Albuquerque exceeded 7,000 cubic feet per second (cfs) in about one of every two years, on average. With the upstream flood-control system, this maximum discharge is now only about 5,400 cfs. During the pre-dam period, peak discharges as high as 42,000 cfs occurred in the Albuquerque reach and 100,000 cfs near San Marcial, causing widespread flooding. The highest peak flow in the Albuquerque reach since completion of Cochiti Dam was 9,500 cfs in 1984, and the maximum flows rarely exceed 8,000 cfs.

In the Albuquerque reach, flow begins to spill into the bosque at 5,000 to 6,000 cfs, which is consistent with the current typical annual peak flow. Under native conditions, the in-channel capacity was much more variable both spatially and temporally than today due to channel enlargement during infrequent high flows and re-adjustment during longer duration low- to-moderate flow periods. The active channel now is more consistent in size along the reach and varies less in size over time than under native (pre-European settlement) conditions due to dampening of the extreme flows by the flood-control system and lateral controls on channel widening by the jetty jacks and infrastructure.

Both the gradient and sediment size tend to decrease in the downstream direction in most rivers, and the Rio Grande is no exception. The Rio Grande falls about 12 feet per mile between Velarde and the mouth of the Rio Chama, decreasing to about 4 feet per mile through the Albuquerque reach and to around 3 feet per mile between El Paso and Fort Quitman in the lower reach. Similarly, the typical bed material upstream from Cochiti Reservoir is mostly gravel and cobbles, changing to mostly sand at and downstream

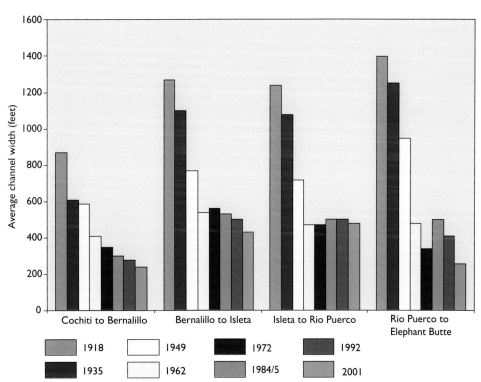

Average width of the active channel of the Rio Grande at different periods, from 1918 to 2001 (Data from the Bureau of Reclamation).

from Albuquerque. Even the sand decreases in size moving downstream from Albuquerque toward Elephant Butte. The interaction of these particles with the flow significantly affects the dynamic behavior of the river. In gravel and cobble bed reaches, relatively high flows are required to move the material and adjust the shape and elevation of the channel bed; thus, the bed is static most of the time. In contrast, flows of nearly all magnitudes can move sand-sized particles; thus, the channel bed is under a state of constant adjustment in these reaches.

Historically, the middle and lower reaches of the Rio

Oblique aerial view of the Rio Grande in the early 1960s, soon after completion of a jetty jack field and dredging of a pilot channel.

Grande carried one of the highest sediment loads of any river in the world, and much of this sediment consisted of very fine silt- and clay-sized particles. Trapping of sediment by Cochiti and other upstream dams has significantly reduced the concentrations. Upstream from Albuquerque, concentrations are now only 2.5 to 3 percent of their historic values. Inflows from local tributaries with fine soils in their watersheds cause the concentrations to increase as one moves downstream from Cochiti toward Elephant Butte.

The fine material had a profound effect on the dynamic behavior of the river under historic conditions, and this effect continues today. During flooding, the silts and clays were carried into the overbanks where they settled out onto the valley floor, building up the floodplain and river banks with cohesive, difficult-to-erode sediment. As a result, the river banks are relatively erosion resistant, limiting the ability of the river to erode laterally. Because the fine sediment tends to have high water-holding capacity and is rich in nutrients, the overbanks provided excellent habitat for riparian vegetation that, historically, consisted of a variety of species including cottonwood. These characteristics are also very suitable for the more recently introduced non-native species such as salt cedar; thus, they are both a blessing and a curse with respect to the health of the riparian zone.

Under native conditions, the Rio Grande through the middle valley was relatively wide and braided with multiple smaller channels separated by active sandbars. In the early 1900s the active channel averaged more than 1,200 feet wide between Bernalillo and the mouth of the Rio Puerco. The river subsequently narrowed to only about one-third of its historic width due to the combined effects of the jetty-jack fields, other human encroachments, and reductions in the peak flows. The river has continued to narrow in some locations since the 1960s, but at a much slower rate, and it is likely approaching equilibrium with the regulated flow regime.

The river has undergone other important changes due to human activities that include changes in size of the bed material and changes in the elevation of the channel bed relative to the floodplain. The reach downstream from Cochiti Dam, for example, has exhibited the classic response to sediment trapping in a reservoir by coarsening the bed material and downcutting as the river made up for the deficit in sediment supply by mining material from the bed. Before construction of the dam, the bed material between Cochiti and Angostura was mostly sand, with 20 to 30 percent medium-sized gravel. The sand has now been largely depleted, and the bed material is primarily gravel. The bed has also downcut by an average of 2 to 4 feet. The gradient restoration facilities that were constructed over the past several years on and near the Santa Ana Pueblo below Angostura are an attempt to mitigate the effects of this downcutting. The opposite effect has occurred at the head of Elephant Butte Reservoir, where the flatter river gradient due to the reservoir has caused sediment deposition, raising the river bed by several tens of feet. The limited hydraulic capacity of the San Marcial Railroad Bridge is an obvious manifestation of this process.

The water development system in the Rio Grande basin is critically important in sustaining water supply and protecting public safety, but this system has changed the physical characteristics and dynamic behavior of the river. Although many of these changes provide positive benefits to our ability to use the river as a resource, they have also had unintended and undesirable consequences to both the health of the ecosystem and our ability to protect critical infrastructure. River managers face a difficult challenge in balancing the costs and benefits of maintaining the system, while finding ways to mitigate the unintended and undesirable effects.

The Surface Water/Ground Water Connection

Robert S. Bowman, *New Mexico Institute of Mining and Technology*

The surface water that we see flowing in the Rio Grande, in irrigation canals, and in drains is intimately linked to the underlying ground water. The surface water and the ground water in fact form one integrated system.

The ground water is not some mysterious lake or stream that's flowing below us out of sight; instead, ground water is simply water held in the spaces between grains of sand, clay, or gravel. In the soil, some of the empty spaces are also filled with air; deeper, the air is displaced, and the pores are totally water-filled. This is the ground water zone. If you dig a hole along the banks of the Rio Grande the hole will remain open and air-filled until a certain depth. Then, as you dig the hole deeper, it will fill with water to some level below the ground surface. This water surface is the level of the ground water table. Below this, you're in the ground water aquifer.

The Rio Grande valley between San Acacia and Elephant Butte Reservoir can be envisioned as a sand- and gravel-filled bathtub with a shallow trench running down the middle. The sides of the bathtub are the uplands and mountains rising to the east and west of the river; the bottom of the bathtub is formed by deep, compacted sediments. At the north end of the tub water flows into the trench (the Rio Grande) through the diversion dam at San Acacia. At the south end the Rio Grande flows into Elephant Butte Reservoir and eventually "drains" from the tub through Elephant Butte Dam into the Mesilla Valley.

As water flows down the Rio Grande some of it seeps through the river bed and spreads out underground. But it can only seep so far before it hits the relatively impermeable sides and bottom of the tub. So the bathtub gradually fills with water, the sand and gravel become saturated from the bottom up, and the water table rises. Once the water table reaches the base of the trench the seepage is reduced, and more of the water entering the valley at San Acacia stays in the river bed all the way to Elephant Butte.

Of course, the bathtub model is an oversimplification of the actual system; some river water can leak out the sides and bottom of the tub, and some water enters the tub from sources (such as rising geothermal waters) other than the river. But in general the water that enters the valley stays in the valley until it reaches Elephant Butte. One exception, however, is the water consumed by evapotranspiration.

Evapotranspiration is the water lost to the atmosphere by evaporation from the soil and open water, and by transpiration of water through the leaves of plants. Another term for evapotranspiration is "consumptive use"—the water that is "consumed" and no longer available in the surface water/ground water system. To a first approximation, the only water that leaves the valley between San Acacia and Elephant Butte is that which is evapotranspired.

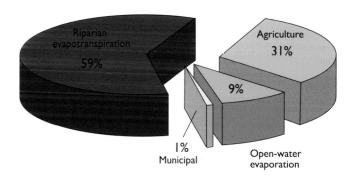

Estimated average annual consumptive use of water between San Acacia and San Marcial. Data provided by Nabil Shafike of the New Mexico Interstate Stream Commission.

The pie chart on this page shows the average annual consumptive use of water between San Acacia and San Marcial, upstream from where the Rio Grande enters Elephant Butte Reservoir. More than half of the consumptive use in this reach is evapotranspiration by riparian vegetation along the river, and almost a third is evapotranspiration from agricultural crops. In total, about 10 percent of the water flowing into the valley at San Acacia is lost to evapotranspiration by the time the river reaches San Marcial. Since most of the dissolved salts remain in the water during evapotranspiration, the salt concentration in the river and in the ground water generally increases as you move downstream. Small inputs of high-salinity water from natural sources other than the Rio Grande also make significant additions to the salt load of the system.

Until recent times the Rio Grande was the only

trench running the length of the valley. However, in response to the drought of the early 1950s and the precipitous drop in Elephant Butte Reservoir, the Low Flow Conveyance Channel (LFCC) was constructed to reduce seepage losses as water moved down the valley. After 1959 some or all of the water from the Rio Grande was diverted to the LFCC at San Acacia Dam and traveled approximately 75 miles to the head of the reservoir.

During the wet period of the mid-1980s Elephant Butte Reservoir filled to capacity; by 1987 there was no need to divert water from the Rio Grande into the flow underground toward it from the east and the west. The LFCC thus currently acts as a surface drain.

The approximately 90 percent of the water remaining in the valley after evapotranspiration losses can move interchangeably between the surface and ground water systems. As mentioned above, water generally seeps from the Rio Grande to the shallow aquifer as it traverses the valley from north to south. But this ground water can appear once again as surface water in the LFCC. This surface water-to-ground water-to-surface water interchange is shown schematically in

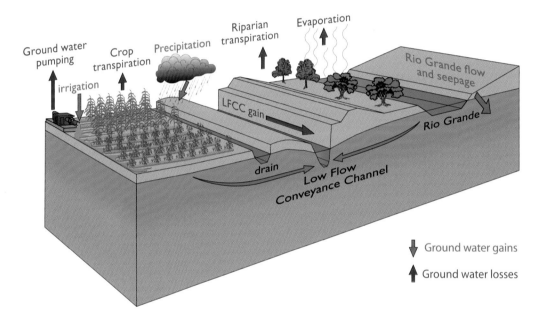

Schematic cross section of the Rio Grande valley showing the linked surface water/shallow ground water system, emphasizing gains and losses to the ground water. The LFCC gain is, in fact, a loss to the ground water.

LFCC. Elephant Butte receded during the 1990s, but sedimentation during the wet years plugged the channel and left the LFCC disconnected from the reservoir. Today the LFCC no longer flows into Elephant Butte but instead discharges into a large "delta" area at the north end of the reservoir, where it supports a dense population of salt cedar and other phreatophytes. Budgetary constraints along with the need to maintain minimum flows in the river (e.g., to support the silvery minnow population) have slowed progress in re-engineering the connection between the LFCC and Elephant Butte Reservoir.

The average bed elevation of the LFCC is below that of the bed of the Rio Grande, and generally lies below the level of the ground water table—it represents a new, deeper trench in the bathtub. Since the LFCC is the topographic low point in the valley, water tends to the illustration on this page. Just as water spilled on a sloping desk runs toward the bottom of the desk, water flows underground from a high point on the water table to points where the water table is lower. Ground water tends to flow from below the Rio Grande west into the LFCC, and from below agricultural fields east into the LFCC. Thus, even though the Rio Grande is no longer directly diverted into the LFCC, much of the river's flow still ends up as surface water flowing toward Elephant Butte via this large drain.

The illustration above also points out many of the other important interactions between surface water and ground water, with an emphasis on gains and losses to the ground water. Water seeps from the Rio Grande as it flows southward, recharging the shallow ground water aquifer. Some of this water reemerges as surface water in the LFCC. Water is diverted from the river at San Acacia Dam to supply the valley's farm irrigation system. A portion of the irrigation water applied to the fields percolates through the soil to recharge the ground water, and can move underground to the LFCC to reappear as surface water. Some of the rain and snow that fall on the watershed can percolate through the soil and reach the water table. And some ground water flows into the valley from the Albuquerque Basin to the north.

Simultaneously water is being lost from the system, and hence from the ground water, due to evapotranspiration. There is direct evaporation from the open water surfaces of the Rio Grande, the LFCC, and agricultural canals and drains. Close to the river, riparian vegetation transpires water to the atmosphere; farther from the river, in the irrigated portion of the valley, crops also transpire water. Water pumped by wells for irrigation and for domestic purposes removes water from the aquifer; much of this water is subsequently lost from the system as evapotranspiration.

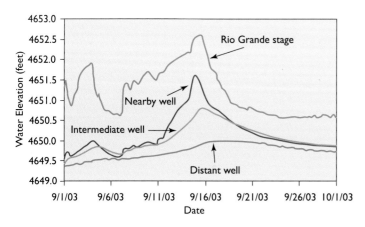

Changing water table levels over a one-month period in wells in response to increasing flows in the Rio Grande below San Acacia Dam. The nearby well is 102 feet from the river, the intermediate well is 388 feet from the river, and the distant well is 1,224 feet from the river.

Compared to the hundreds of thousands of acre-feet that flow down the Rio Grande between San Acacia and Elephant Butte in an average year, relatively little ground water is pumped from the aquifer. During the wet period from the early 1970s to the late 1990s, when surface water was plentiful, almost no water was pumped for irrigation. Even during the dry years of the late 1990s and early 2000s, most of the water for irrigation was provided by surface water diversions from the river. Domestic well pumping in the region amounts to only a few hundred acre-feet of water per year.

The ground water pumping situation in the San Acacia reach is in sharp contrast to that of Albuquerque and Bernalillo County. In 2005 about 100,000 acre-feet of water were pumped by Albuquerque city wells, and thousands more were pumped by domestic wells.

The ground water table responds very rapidly to changes in river flow, and flow in the river likewise responds quickly to changes in the elevation of the water table. The strong connection between surface and ground water levels is shown in the graph on this page, which illustrates the water table response to increasing river flows below San Acacia Dam following a series of heavy rainfalls in the fall of 2003. The changing river flow is shown by the stage (or height) of water in the river relative to sea level. The water table elevations in wells at different distances away from the river are also shown relative to sea level.

The ground water response to fluctuating river flows is dramatic. Within a few hours after an increase in river stage, increased seepage from the river causes the water table at a well very near the river to rise. An intermediate well responds later, and the well most distant from the river takes several days to fully respond. As the flow in the river decreases and the river stage drops, the water table drops correspondingly. In a sense the aquifer "breathes in" water in response to increased flow in the river. When the river flow decreases, the aquifer "breathes out" and returns some of the water back to the river.

A related but contrasting situation occurs when ground water is pumped from the aquifer by irrigation or domestic wells. Pumping of wells, particularly when they are close to the river, causes a drop in the water table elevation and an increase in the gradient between the river and the water table. Analogous to the gradient between the Rio Grande and the LFCC, this increased gradient causes more water to seep out of the river bed, reducing the flow in the river. The effect of a single pumping well on river flow may be difficult to detect, but the combined pumping of many wells can have a measurable effect. The large-scale pumping of the aquifer in the Albuquerque area since the 1950s has caused the water table to drop hundreds of feet in some areas, with a resultant increase in seepage from the river. This extraction of ground water has far exceeded the ability of river seepage to replenish it.

As shown by the above examples, the strong interconnections between surface water and ground water mean any perturbation in one part of the system affects the other. Ground water pumping lowers the water table and increases river seepage, ultimately reducing flow in the river. Reducing seepage (e.g., by concrete-lining of irrigation canals or of the river bed itself) reduces recharge to the aquifer and allows the water table to drop, potentially reducing riparian evapotranspiration to the point that undesirable species such as salt cedar as well as desirable species such as cottonwood may be unable to survive. Thus, any alteration in river management will affect ground water dynamics, just as increased ground water pumping, particularly if water is exported out of the immediate vicinity, will cause changes in the Rio Grande and the riparian community it supports.

Ecology of the Middle Rio Grande of New Mexico

Mary J. Harner and Clifford N. Dahm, *University of New Mexico*

The Middle Rio Grande is located in central New Mexico between Cochiti Dam and Elephant Butte Dam. People have used the Middle Rio Grande and its surrounding land for centuries for agriculture, grazing, and timber. However, widespread physical alterations to the river did not occur until recent times. Major changes took place during the twentieth century as people became concerned about floods, accumulation of salt in agricultural soils, and delivery of water downstream to Texas and Mexico. Flood-control dams, levees, and diversion structures were built, included patches of forests of various ages (ranging from recently established to mature), shrubs, herbaceous plants, and grasslands. Islands in the river channel contained a mix of semi-aquatic habitats and early successional vegetation. This diversity of habitats over small areas provided a mix of resources for plants, animals, and microbial communities. In addition, the river flooded predominantly in spring, as snow melted from mountains in northern New Mexico and southern Colorado, and in late summer during monsoon-season storms. These floods increased soil moisture, cleared

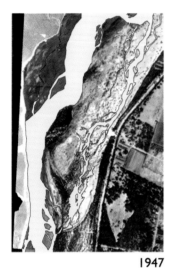
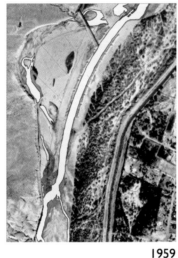
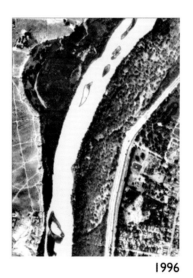

1947　　　　　**1959**　　　　　**1996**

Aerial photos highlighting the channels of the Rio Grande in Albuquerque near the Rio Grande Nature Center. Photos depict the river in 1947, prior to regulation, with overbank flooding (note colonization of young cottonwoods in vicinity of side channel); in 1959, following the 1950s drought and immediately after the construction of levees (which parallel the channel) and jetty jacks (perpendicular to the channel); and in 1996, with a simplified river channel and a contiguous, mixed forest of cottonwood and non-native species. Photos depict a 2-km stretch of river.

including Elephant Butte Dam (1916), drainage ditches parallel to the river channel (1920s), levees (1950s), and Cochiti Dam (early 1970s). In addition, urban development extended closer to the river.

Before regulation the Middle Rio Grande flowed through a network of channels separated by islands. Channels changed position frequently over the river's sandy foundation, and this movement maintained a diversity of habitats. Aquatic habitats included the main channel, side channels of shallow, slowly moving water, as well as ponds and marshes. Terrestrial habitats vegetation from river banks, deposited sediment from upstream, altered channel structure, decomposed organic matter, and distributed aquatic organisms and seeds.

Regulation disconnected the river from its floodplain. Installation of levees constrained the river to a single floodway, and side-channels, wetlands, and ponds nearly disappeared. Dams eliminated large floods, increased flows in the river during summer, trapped sediment, and produced barriers to movement of aquatic organisms. Without floods, dense forests developed along the river because scouring flows were unavailable to remove vegetation. The dominance of non-native plants increased throughout the riparian areas. Wood and leaves accumulated on the forest floor because water was unavailable to decompose and transport it from the forest. The combination of dense vegetation, large accumulations of wood and decaying leaves, and lack of wet soil increased the size and frequency of wildfires. Continued growth of the human population led to more agricultural and urban floodplain development, increased inputs of treated waste water to the river, and increased use of river and ground water.

VEGETATION

Riparian forests in low-lying regions along the Middle Rio Grande are locally known as "bosque," the Spanish word for woodland. The native bosque contains forests of Rio Grande cottonwoods, with shrubs and herbaceous plants growing beneath the cottonwoods. Native woody shrubs along the Middle Rio Grande include Goodding and coyote willows, New Mexico olive, baccharis, and false indigo bush. Sedges, rushes, cattails, and yerba mansa grow in moist soils. Plants more tolerant of drier and saltier soils, such as mesquite and salt grass, live on higher or disconnected surfaces.

Cottonwood trees require floods to establish. They release windblown seeds in spring at the time when the river naturally flooded. Cottonwood seedlings need direct sunlight, and their roots must touch wet soil. Floods create sites for recruitment of cottonwoods by scouring away plants that would otherwise shade young seedlings and by elevating soil moisture. Ideal conditions for cottonwood establishment occurred historically once every 5 to 10 years. These optimal conditions no longer exist. Rio Grande cottonwoods have not reestablished themselves over large areas since the early 1940s, following the last large floods. Therefore, most Rio Grande cottonwoods are 60 years old or older. Young trees are not replacing mature cottonwoods because large floods have been eliminated.

Non-native plants that do not require such precise conditions for recruitment are replacing cottonwoods along the Middle Rio Grande. Salt cedar and Russian olive are common non-native trees in the bosque. They release seeds throughout the summer and can grow under the shade of other vegetation. Whether salt cedar uses more water than native plants or increases salt in soil is a focal area of study. Measurements of water loss along the Middle Rio Grande show that forests containing a mixture of cottonwoods and non-native plants use the most water. Forests of only cottonwood, only dense salt cedar, or only Russian olive use slightly less water than cottonwood forests with a non-native understory. People are removing non-native plants over large areas to possibly conserve water, reduce the risk of wildfires, and encourage growth of native plants.

MICROBIAL COMMUNITIES

Microorganisms, such as bacteria and fungi, play critical roles in ecosystems. They decompose organic matter, such as leaves and wood, and transform nutrients into forms that are available for uptake by algae and plants. Flooding stimulates microbial activity, and abundances of bacteria and fungi are higher at sites that regularly flood. Some fungi, known as mycorrhizal fungi, live on or within the roots of plants. These fungi provide nutrients to plants in exchange for energy (carbon) from plants. Roles of soil fungi in riparian ecosystems are a current area of research along the Middle Rio Grande. Some researchers seek to understand whether fungi and other microorganisms should be added to soil to promote plant growth during restoration projects aimed at reestablishing native vegetation following disturbances, such as fire.

INVERTEBRATES

Hundreds of species of invertebrates inhabit the Middle Rio Grande. Aquatic invertebrates include various species of mayflies, stoneflies, caddisflies, midges, and true flies. Aquatic invertebrates use leaf and woody debris for habitat, and they commonly live in shallow, low-velocity channel edges and backwaters. Today, lack of habitat limits the abundance of some aquatic invertebrates in the Middle Rio Grande. In addition, degradation of water quality negatively affects some species. For example, several species of mollusks inhabit isolated springs along the Middle Rio Grande. They are sensitive to changes in their environment, and several species are now listed as endangered. In contrast, the introduced Asiatic clam, which can tolerate degraded conditions in the river, is now found throughout the Middle Rio Grande.

Terrestrial invertebrates also have important roles in the bosque ecosystem. Some terrestrial invertebrates break down organic matter by chewing it into pieces. Two non-native terrestrial isopods (pill bugs) are dominant decomposers of organic matter in these forests. Other terrestrial invertebrates feed upon leaves in the canopy of trees, and some prefer stressed cottonwoods. Flooding has been shown to affect the composition of invertebrates. Crickets, a native decomposer, tend to increase in abundance when soil moisture rises. Abundance of carabid beetles increases at sites that continue to flood, and their densities might serve as indictors of hydrologic connectivity between the river and the forest.

FISH

Historically the Rio Grande in New Mexico contained 17 to 27 species of fish, including big river fishes such as longnose gar, shovelnose sturgeon, and American eel. Reduced river flows and increased sedimentation led to the extirpation of many fish. Elephant Butte Dam stopped upstream migration of large species. Operation of Cochiti Dam reduced water temperatures, sediment loads, and habitat complexity throughout the Middle Rio Grande. During recent years, drying of the main river channel killed fish and reduced their migrations. Many native fish have been lost from Rio Grande, and 13 to 19 non-native species of fish have been introduced, including the common carp and white sucker. Only one native minnow species remains, the Rio Grande silvery minnow. It once lived throughout the Upper and Lower Rio Grande, as well as the Pecos River basin. Today the silvery minnow lives only between Cochiti Dam and Elephant Butte Reservoir.

AMPHIBIANS & REPTILES

Amphibians, which live part of their life cycle in water, once thrived in wet meadows, marshes, and floodplain ponds along the Rio Grande. Amphibians living along the Middle Rio Grande include: Couch's spadefoot toad, Woodhouse's toad, great plains toad, and northern leopard frog. Some native amphibians, especially the northern leopard frog, have been negatively affected by reductions in availability of floodplain pools, as well as by predation by introduced bullfrogs. Common reptiles that live along the Middle Rio Grande include the eastern fence lizard, New Mexico whiptail lizard, spiny softshell turtle, and common garter snake.

MAMMALS

Several species of large mammals, including grizzly bears, jaguars, and gray wolves once inhabited the Rio Grande valley but are no longer present. People also depleted populations of beavers during the nineteenth century. Beavers were restocked from 1947 to 1958 and now maintain healthy populations in riverbanks. Rock squirrels and valley pocket gophers also live along the Middle Rio Grande. Pocket gophers burrow in floodplain soils, and their activities move deep soil to the surface, thus increasing the cycling of nutrients and movement of soil microbes. Small mammals along the Middle Rio Grande include the white-footed mouse, house mouse, tawny-bellied cotton rat, western harvest mouse, hispid cotton rat, white-throated woodrat, Ord's kangaroo rat, and piñon mouse. Many small mammals prefer grassy areas of the floodplain. The white-footed mouse, which often nests in cavities of trees, avoids drowning during floods by climbing trees. Three species of bats commonly roost under old wooden bridges along the Middle Rio Grande. These bats are primary consumers of night-flying insects.

BIRDS

Riparian forests in the desert Southwest provide nesting and foraging habitats for resident and migratory birds. Riparian areas often have high numbers of avian species, and this has been documented in woodlands and marshes along several rivers. More than 270 species of birds use habitats along the Rio Grande, and many breed along the river. Birds of the Middle Rio Grande include Bewick's wren, great blue heron, black-chinned hummingbird, white-crowned sparrow, downy woodpecker, and great horned owl. Tens of thousands of waterfowl, such as snow geese and sandhill cranes, spend the winter along the Middle Rio Grande, especially in the areas between Bernardo and the Bosque del Apache National Wildlife Refuge. Many of these birds feed on alfalfa and corn throughout the river valley. Birds bring nutrients from adjacent terrestrial areas to wetlands at night while they roost in the safety of shallow water or islands. Chemical signals of corn and alfalfa, which can be identified by stable isotopes, appear throughout the food web, especially in fish and crayfish.

Habitat loss via reduction of wetlands and decreases in forests with multiple layers of vegetation has contributed to declines in some species of birds along the Rio Grande. Birds use a variety of riparian habitats for nests. Some build nests on the ground or in shrubs, others use the canopy of trees, and some nest within cavities of trees. The endangered southwestern willow flycatcher relies upon dense shrubs along waterways for nests. Introduced European starlings, which nest in tree cavities throughout the bosque, compete with native cavity nesting birds. Bird use of native versus non-native plants along the Middle Rio Grande is an area of current research. Some research suggests that riparian forests with a mix of native trees and shrubs of different sizes have the greatest diversity of birds.

PEOPLE

The Rio Grande and its bosque have provided resources for humans for thousands of years. Native Americans living in pueblos used water from the Middle Rio Grande for irrigation. During the seventeenth century, Spanish settlers developed a permanent system of diversions known as acequias. European settlers used riparian areas for cattle grazing and timber harvesting. Today more than half of New Mexico's population resides in the Rio Grande basin. Managers seek to balance human demands on the river with needs of other organisms that rely upon the Middle Rio Grande. Efforts are underway to restore parts of the bosque to native vegetation and to reduce risks of catastrophic fires. Managed floods that match the timing, but only a fraction of the size, of historic floods have been used in the Middle Rio Grande to demonstrate the importance of floods to the ecosystem. General knowledge of bosque ecology is reaching the public through programs like the Bosque Ecosystem Monitoring Program, a program that brings hundreds of school children to the river each year to learn about and measure components of the ecosystem.

Suggested Reading

Middle Rio Grande ecosystem: Bosque Management Plan, Crawford, C. S., Cully, A. C., Leutheuser, R., Sifuentes, M. S., White, L. H., and Wilber, M. P., U.S. Fish and Wildlife Service, District 2, Albuquerque, NM, 1993.

Middle Rio Grande biological survey, Hink, V. C. and Ohmart, R. D., Report submitted to U.S. Army Corps of Engineers, Albuquerque, NM, 1984.

Managing Surface Waters on the Upper Rio Grande

Rolf Schmidt-Petersen, *Rio Grande Bureau, New Mexico Interstate Stream Commission*

The Upper Rio Grande basin stretches from the headwaters of the Rio Grande in Colorado to Fort Quitman, Texas, 70 miles southeast of El Paso. Surface water flow in the basin is highly variable and can easily vary 50 percent above or below the long-term mean flow. Surface water management in the Upper Rio Grande basin has evolved to address this variability so that water users can be protected in times of both drought and flooding. People sought to fund and build water projects that would store floodwater, thus reducing flooding risk and allowing the stored floodwater to be used later in the year, when natural flows were low, to irrigate crops. One significant example, the construction and subsequent operation of the Rio Grande Project, resulted from controversy associated with a multi-year drought in the 1890s.

Several years of low snowmelt, in combination with increased irrigation diversion, mostly in the San Luis Valley of Colorado, resulted in water shortages throughout the basin with the most pronounced effects experienced in the lower parts of the basin. Texan, New Mexican, and Mexican farmers along the international border suffered significant shortages in supply. Although individual farmers complained for years, it was not until Mexico formally complained that the U.S. government became actively involved. In 1896, in an attempt to "freeze" development upstream of the border, the Secretary of the Interior declared an embargo prohibiting use of federal funds or grants of easements across federal land for water development projects. That action effectively stopped additional large-scale water development upstream of what is now Elephant Butte Reservoir until the 1920s.

In 1906 the United States and Republic of Mexico resolved their differences when they entered into the International Treaty of 1906. Except in times of extraordinary drought, the treaty guarantees Mexico 60,000 acre-feet of water each year at El Paso. The U.S. Congress authorized construction of the Rio Grande Project in part to assure the guarantee could be fulfilled.

THE RIO GRANDE PROJECT

The Rio Grande Project, located in southern New Mexico and northwest Texas, consists of two reservoirs (Elephant Butte being one) and four river diversion dams. It extends 130 miles south from Elephant Butte Reservoir past Las Cruces, New Mexico, and El Paso, Texas, to the Hudspeth County line in Texas. It was constructed by the U.S. Bureau of Reclamation in the early twentieth century, in part to comply with terms of the 1906 treaty. Its primary purpose is to deliver water to 160,000 acres of land in New Mexico and Texas for irrigation, and to provide 60,000 acre-feet of water annually to Mexico.

Construction and operation of the Rio Grande Project resolved issues between the U.S. and Mexico about surface water delivery north of Fort Quitman, Texas, and south of Elephant Butte Reservoir. It established the infrastructure and operations necessary for the U.S. to store Rio Grande waters in Elephant Butte Reservoir and to deliver the stored water for irrigation use in New Mexico, Texas, and Mexico. The project also significantly reduced flooding risk and improved surface water security for people living south of Elephant Butte Reservoir. Although the amount of water available from the Rio Grande Project has varied through time, the project clearly has brought much more certainty to landowners downstream.

Elephant Butte Reservoir was the first large storage reservoir on the Rio Grande and is the primary storage reservoir for the Rio Grande Project. It is the largest reservoir in the Upper Rio Grande basin. Its primary authorized purpose is to provide water for irrigation. Although Congress authorized a "recreation pool" for the reservoir in 1974, no permanent source of water was reserved for the pool. (A recreation pool is storage space within the reservoir for water that would never be released; it would be lost only through evaporation. The idea is to hold some water in the reservoir even during the driest times.) The reservoir therefore has no defined minimum pool, and recreational water users do not have a water right.

The Bureau of Reclamation and two U.S. irrigation districts (the Elephant Butte Irrigation District in New Mexico and the El Paso Water Improvement District No. 1 in Texas) run the Rio Grande Project. The Bureau of Reclamation makes allocations of project water to the two U.S. districts and Mexico during each irrigation season (March through October) through the U.S. International Boundary and Water Commission.

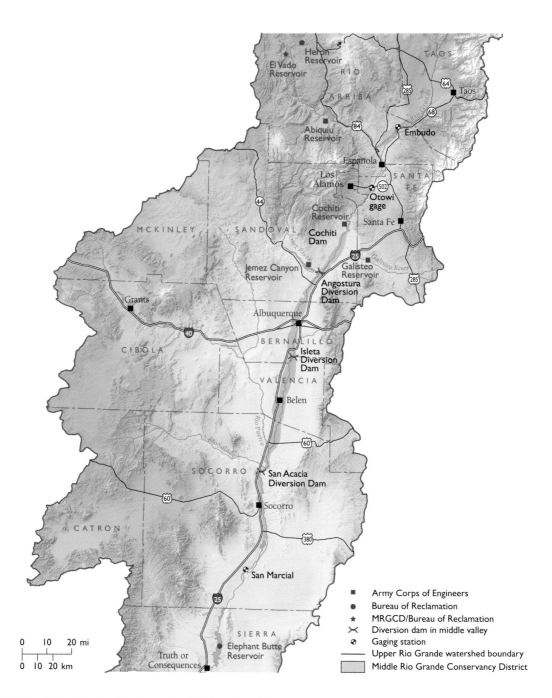

The Rio Grande basin in New Mexico north of Elephant Butte Reservoir.

Each district's board of directors uses the Bureau of Reclamation's allocations to develop general plans for storage releases (of their respective allocations) during the irrigation season. Individual farmers are allocated a set amount of surface water by their district for use during the irrigation season, based upon the irrigable acreage held by the farmer. An individual farmer orders surface water from the district as needed until his/her allotment is fully delivered. Each district and Mexico (through the U.S. Section of the International Boundary and Water Commission) make regular requests of the Bureau of Reclamation for delivery of water at specific diversion dams to meet their farmers' orders. The Bureau of Reclamation then releases from storage the amount of water needed to provide the requested river diversions.

Rio Grande Project water is also used for municipal and industrial purposes. The City of El Paso is a landowner in El Paso Water Improvement District No. 1, has fallowed the land it owns or leases, requests surface water from the district as any district farmer would, and then uses the delivered surface water for municipal and industrial purposes.

The Rio Grande Project set the stage for battles between direct flow users north of Elephant Butte Reservoir and water users south of the reservoir. Water use north of the reservoir continued to be constrained by the ability of individuals to divert water from the river, the variability in supply, sedimentation, the ability of people upstream to take and consume water, and the inability of most people to receive federal funds and permissions to cross federal land to improve access to surface water. In both Colorado and New Mexico above Elephant Butte Reservoir, water users were experiencing shortages especially toward the end of the irrigation season.

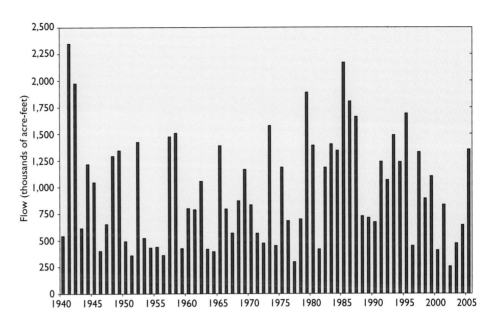

Unregulated native Rio Grande flow at the Otowi gage near Los Alamos. The Otowi gage provides an estimate of annual water supply conditions both upstream and downstream. If upstream water users are experiencing flooding or drought, such events are observed in the gage record. Because 80–85 percent of the water that flows in the Upper Rio Grande basin in most years flows past this point, the gage provides an estimate of how much surface water was available downstream.

CONFLICT AMONG COLORADO, NEW MEXICO, AND TEXAS

By 1925 the Secretary of the Interior had lifted the federal water development embargo of 1896 when he authorized rights of way for construction of a reservoir in Colorado. The embargo had remained in place long after Elephant Butte Reservoir became operational. People downstream of the reservoir were adamant about the need for the embargo to remain in order to protect Elephant Butte Reservoir storage. They had the federal government as an ally, because neither they nor the federal government wanted to allow reservoirs to be built upstream, given that such reservoirs would reduce the amount of water making it to Elephant Butte Reservoir. People upstream were adamant that damage from floods and drought made it difficult for them to take water from the river and use it as they had before the embargo.

Upon partial lifting of the embargo, Coloradans built several small reservoirs without federal funding. After lifting of the embargo, New Mexicans in the middle valley sought to reclaim lands damaged during the preceding 30 years, reduce flood risk, and improve their irrigation infrastructure. The Middle Rio Grande Conservancy District was organized in 1925 to do just that. It drained waterlogged lands and removed some seventy different river diversion points, consolidating them to the four that exist today (Cochiti, Angostura, Isleta, and San Acacia) for delivery of water to district farmers. Some seventy different acequias operating in the middle valley in the early 1920s were subsumed by the conservancy district. It centralized the irrigation delivery system and constructed El Vado Dam and Reservoir.

The efforts of the Middle Rio Grande Conservancy District resulted in reclamation of previously waterlogged lands within the district, significantly improved water delivery to farmers, and somewhat reduced flood threat. Those efforts—and funding—were largely non-federal in origin, with the exception of funds associated with a 1928 act of Congress that provided funding for work associated with water delivery to Pueblo lands within the district. The 1928 act also established broad categories of water rights and priorities within the district. Most explicitly, it designated specific amounts of lands within the six Middle Rio Grande pueblos as having a senior water right to any other Middle Rio Grande Conservancy District lands

How Surface Water Management Decisions are Made

The Colorado Division of Water Resources oversees surface water diversions north of the state line with New Mexico to deliver water to its farmers and to the State of New Mexico under the 1938 compact. The Bureau of Reclamation, U.S. Army Corps of Engineers, U.S. Bureau of Indian Affairs, New Mexico Office of the State Engineer/Interstate Stream Commission, and Middle Rio Grande Conservancy District collaborate on irrigation deliveries, maintaining U.S. Fish and Wildlife Service Biological Opinion river flow targets in the middle valley, and compact management between the state line with Colorado and Elephant Butte Dam. The Bureau of Reclamation, U.S. International Boundary and Water Commission, Elephant Butte Irrigation District, and El Paso Water Improvement District No. 1 collaborate on surface water management for the Rio Grande Project.

The Colorado Division of Water Resources tracks Rio Grande surface water flows at the state line with New Mexico relative to its upstream 1938 compact index flows and associated delivery requirements. Based upon their projections of needed deliveries, the district engineer coordinates with water users to decide when the irrigation season will begin and end and curtails surface water diversions during the irrigation season (sometimes for water users with rights that significantly pre-date the signing of the compact) in order to meet its required annual deliveries to New Mexico. The Corps of Engineers coordinates with the Division of Water Resources and the Bureau of Reclamation to oversee flood operations at Platoro Reservoir on the Conejos River, when downstream river flow conditions warrant such operations.

In the Upper and Middle Rio Grande valleys of New Mexico, surface water management can be separated into three main categories, the winter, the irrigation season, and flood control operations. During the winter (November through February), the Bureau of Reclamation, the Army Corps of Engineers, and the Interstate Stream Commission coordinate to manage the system's reservoirs in compliance with the 1938 compact to route 1938 compact deliveries to Elephant Butte Reservoir and to maintain river flow on the Rio Chama by moving San Juan–Chama Project water to various delivery points. Generally, management for endangered species flow targets is not necessary during the winter.

During the irrigation season (March through October), the Bureau of Reclamation, the Corps of Engineers, the Bureau of Indian Affairs, the Fish and Wildlife Service, the Middle Rio Grande Conservancy District, the Interstate Stream Commission, and the Albuquerque Bernalillo County Water Authority confer daily, as necessary, on river flow conditions throughout the Upper and Middle Rio Grande valleys. Based upon the amount of surface water flowing naturally into the middle valley, weather conditions in both the upper and middle valleys, irrigation demand, and the needed Biological Opinion flows, decisions are made on whose water and how much water to release from upstream reservoirs. The agencies coordinate to deliver water to the Middle Rio Grande Conservancy District diversion dams and maintain compliance with the Fish and Wildlife Service Biological Opinion. During the snowmelt runoff they work, as needed, to provide flows for spawning and recruitment of silvery minnow. During low flow periods they coordinate to observe and manage river drying throughout the Middle Rio Grande valley, salvage silvery minnow, and thus reduce take of silvery minnow.

As warranted by snowpack, reservoir, and river conditions, the corps manages its reservoirs (Abiquiu, Cochiti, Jemez Canyon, and Galisteo) to store floodwater and provide flood damage protection to lands and people living in the upper and middle valleys from Abiquiu to Elephant Butte Reservoir. The water management agencies coordinate with the State Emergency Management Office and individual county emergency managers, as necessary, to provide flood warnings. The Bureau of Reclamation conducts emergency operations as necessary to limit the potential for flooding within the Middle Rio Grande Project. Additionally, the Middle Rio Grande Conservancy District may increase its diversions from the river in order to reduce pressure on the downstream levee system.

In the Lower Rio Grande, the Bureau of Reclamation, the Elephant Butte Irrigation District in New Mexico, and the El Paso Water Improvement District No. 1 in Texas run the Rio Grande Project. The Bureau of Reclamation makes allocations of project water to the two U.S. districts and Mexico during each irrigation season (March through October) through the U.S. Section of the International Boundary and Water Commission. Each district develops general plans for storage releases (of their respective allocations) during the irrigation season and allocates water to individual farmers. An individual farmer orders surface water from his/her district as needed until his/her allotment is fully delivered. The City of El Paso is a landowner within the El Paso Water Improvement District No. 1, receives annual allocations of water from the district, and uses its allotted surface water to provide drinking water to its citizens. Each district and Mexico (through the U.S. Section of the International Boundary and Water Commission) make regular requests of the Bureau of Reclamation for delivery of water at specific diversion dams to meet their farmers' orders. The Bureau of Reclamation then releases from storage the amount of water needed to provide the requested river diversions. The U.S. Section of the International Boundary and Water Commission oversees flood control operations.

and indicated that lands not previously irrigated (newly reclaimed lands) had the most junior priority.

As part of an effort to have the 1896 embargo lifted, the state legislatures of New Mexico and Colorado each passed statutes in 1923 authorizing designation of commissioners to pursue formulation of an interstate compact for the Rio Grande. In 1926 Texas joined the Compact Commission to protect its water user interests.

THE 1929 RIO GRANDE COMPACT—THE "STANDSTILL COMPACT"

The temporary Rio Grande Compact, signed in February 1929, was designed to maintain the status quo in the basin. Major parts of the agreement included:

- A condition that neither New Mexico nor Colorado could increase diversions or storage of water on the Rio Grande until such time as the resulting depletions were offset by drainage projects. The primary assumption is that the drainage projects would return an equal amount of water to the river as that consumed by the reservoir storage and associated release operations. This facilitated efforts in Colorado to construct a drainage project to drain water from the "closed basin" in the San Luis Valley of Colorado to the Rio Grande and build a mainstem reservoir. In New Mexico it facilitated efforts to improve drainage in the middle valley and to construct El Vado Reservoir.

- A provision for creation of a Compact Commission to permanently and equitably apportion the river in the Upper Rio Grande basin.

- A provision that each state will maintain stream flow gaging stations and exchange records of measurements.

The 1929 compact created a path for the three states to agree upon water use limits and delivery requirements. It also set the stage for Coloradans and New Mexicans north of Elephant Butte Reservoir to push aggressively for funds, both private and federal, to improve their ability to control and use water. The period between 1929 and 1938 was one of cooperation and conflict: the Rio Grande Joint Investigation of water uses in the Upper Rio Grande basin was being conducted collaboratively in the midst of a U.S. Supreme Court lawsuit filed by Texas against New Mexico and the Middle Rio Grande Conservancy District, in part over the construction of El Vado Dam and Reservoir. Not many people realize that the Rio Grande Compact of 1938 settled a U.S. Supreme Court lawsuit in addition to resolving water uses from the headwaters of the river in Colorado to Fort Quitman, Texas; the lawsuit was dismissed upon signing of the 1938 compact. And the 1938 compact was signed shortly after agreements between the U.S. government and water users in Elephant Butte Irrigation District and El Paso Water Improvement District No. 1 were signed, dividing the irrigable lands of the Rio Grande Project, with 57 percent going to Elephant Butte Irrigation District and 43 percent to El Paso Water Improvement District No. 1.

THE 1938 RIO GRANDE COMPACT

The 1938 compact was designed to stabilize water depletions in the Upper Rio Grande basin as they existed in 1929. It reflects the efforts of the negotiators to ensure that the same general quantity of flow that made it to Elephant Butte Reservoir before 1929 continued to do so. The 1938 Rio Grande Compact established surface water delivery obligations for Colorado to New Mexico near the state line and New Mexico to Texas at San Marcial near the headwaters of Elephant Butte Reservoir. The compact sets annual delivery obligations of Colorado to New Mexico and a nine-month delivery obligation of New Mexico to Texas. As a result, it also established depletion amounts for both Colorado and New Mexico of the natural flow of the river at certain points.

The 1938 compact provides for both drought and high-flow conditions. It does not require Colorado or New Mexico to deliver the exact amount of water scheduled annually each and every year. The compact allows for the accumulation of over deliveries (credit) and under deliveries (debit). It allows for significant debits in deliveries to accrue before either Colorado or New Mexico are in violation of the compact. Additionally, if Elephant Butte Reservoir is full and spills, the compact provides that all credits or debits are wiped out.

What does the 1938 Rio Grande Compact mean in practice for people living in New Mexico or Colorado north of Elephant Butte Reservoir? The Rio Grande Compact establishes depletion amounts for Colorado and New Mexico of the natural flow of the river at certain points. Although it is up to each state to decide how its water is used, any new use has to be balanced by reduction of an existing use. Alternately, new uses can be supported using imported water supplies such as San Juan–Chama Project water. Ground water can be used as long as the impact of that use on the Rio Grande is offset. Typically, ground water withdrawn will eventually deplete the river in the same amount as that pumped. Furthermore, the compact places restrictions on the operation of reservoirs in New Mexico and Colorado constructed after 1929 if the storage supply of the Rio Grande Project drops below a specified level, or if either state has an accrued compact debit.

THE 1948 RIO GRANDE COMPACT RESOLUTION

In 1948 the Rio Grande Compact Commission approved a resolution moving the New Mexico delivery point from the headwaters of Elephant Butte Reservoir to Elephant Butte Dam. The move was made because of difficulty measuring the flow of the Rio Grande at the old delivery point and to develop an annual delivery obligation of New Mexico to Texas. The commission did so in part by estimating long-term evaporation rates from Elephant Butte Reservoir, based upon historic Rio Grande Project operations.

With the above changes, the negotiators also removed a clause from Article IV of the 1938 compact concerning application of New Mexico's delivery schedule. The clause had required changes to the delivery schedule should New Mexico deplete the runoff of tributaries to the Rio Grande between Otowi Bridge and San Marcial during the summer months by works constructed after 1937. The resolution removed one impediment to the federal government's construction of Middle Rio Grande Project facilities. Finally, because many of the depletions in the middle valley

A Few Definitions

Direct Flow Right—A water right, with a priority date, to divert a certain amount of water from a stream and put it to beneficial use.

Storage Right—A water right to store surface water at times when downstream senior direct flow and storage rights are satisfied. The stored water is released upon the call of the storage right holder for downstream beneficial use. Once released, the storage water suffers natural losses between the point of release and point of diversion.

Rio Grande Project—A U.S. Bureau of Reclamation water project in southern New Mexico and northwest Texas consisting of two reservoirs and four river diversion dams. The project extends 130 miles south from Elephant Butte Reservoir past Las Cruces, New Mexico, and El Paso, Texas, to the Hudspeth County line in Texas. The project was constructed to deliver water to 160,000 acres of land in New Mexico and Texas for irrigation purposes and to provide 60,000 acre-feet of water to Mexico annually.

Middle Rio Grande Conservancy District—An entity organized under the New Mexico Conservancy Act of 1923, as amended to plan for reclamation, flood protection and irrigation in the Middle Rio Grande. The district is located in the middle of the state, extending south from Cochiti Reservoir 150 miles past Albuquerque and Socorro to the northern boundary of the Bosque del Apache National Wildlife Refuge. The district oversees operation of a reservoir on the Rio Chama and four river diversion structures within the Middle Rio Grande valley.

Elephant Butte Irrigation District—An entity governed by an elected board and organized in New Mexico in 1918 to contract with the Bureau of Reclamation for irrigation works ultimately servicing some 90,000 acres of irrigable land within the New Mexico portion of the Rio Grande Project. The district requests reservoir releases from the Bureau of Reclamation for New Mexico farmers, diverts that water at one of three diversion dams, and then delivers the water to its constituent farmers.

Middle Rio Grande Project—A U.S. Army Corps of Engineers and Bureau of Reclamation project authorized in 1948 and 1950 to provide additional flood control, storage, channel rectification, restoration of irrigation works, and other efforts on the Rio Grande between Velarde, New Mexico and Elephant Butte Reservoir.

San Juan–Chama Project—A U.S. Bureau of Reclamation project consisting of three diversion dams, three tunnels, and one reservoir. The project is used to deliver a portion of New Mexico's Upper Colorado River Compact water apportionment from the San Juan Basin to the Rio Grande Basin. The Bureau of Reclamation diverts water from three tributaries of the San Juan River in southwest Colorado and transports it under the continental divide via a series of tunnels to Heron Reservoir on Willow Creek just above its confluence with the Rio Chama. The bureau contracts with various entities in the Rio Grande basin north of Elephant Butte Reservoir for annual deliveries of San Juan–Chama Project water from Heron Reservoir.

are natural, the resolution established a need for New Mexico to maintain the river through the middle valley to both control natural depletions and efficiently deliver water to Elephant Butte Reservoir. To put it another way: In order for New Mexico to increase its human water use in the middle valley it must reduce and control existing natural uses (evapotranspiration from the bosque or evaporation from open water).

THE MIDDLE RIO GRANDE PROJECT (1948 AND 1950 FLOOD CONTROL ACTS)

Large floods in the early 1940s resulted in significant short-term and long-term harm for people living in the middle valley. The Middle Rio Grande Conservancy District's irrigation infrastructure suffered significant damage; the district was nearly bankrupt, lands in the valley became salinated or waterlogged, and the river channel ceased to exist south of the Bosque del Apache National Wildlife Refuge. Given the hardship, residents once again sought to reduce flood risk and improve conveyance of water. Additionally, New Mexico began to accrue significant compact under deliveries.

The Middle Rio Grande Project, a joint U.S. Army Corps of Engineers and U.S. Bureau of Reclamation project, was therefore advocated and supported by many parties as a way of addressing the myriad of middle valley water problems. The project included construction of four large flood control reservoirs, removal of multiple miles of river channel from the valley, construction of the Rio Grande "floodway" and the Low Flow Conveyance Channel, and reconstruction of parts of the Middle Rio Grande Conservancy District.

The operations of all the Corps of Engineers flood control reservoirs must comply with the 1938 compact. The corps cannot store native Rio Grande water except for floodwater, cannot deviate from defined operations without approval of the Compact Commission, and must pass floodwater through the system at the highest "safe" rate possible. Under certain circumstances, the corps cannot release stored floodwater after July 1 of any year until the end of the Middle Rio Grande Conservancy District's irrigation season.

The river realignment and water conveyance facilities of the Middle Rio Grande Project reduced water consumption and aided New Mexico in meeting its delivery obligations. The Middle Rio Grande Project was and remains a key element in New Mexico's ability to maintain compact compliance. Consequently, maintenance of the Middle Rio Grande Project is vital for the state.

THE SAN JUAN–CHAMA PROJECT

This project, constructed by the U.S. Bureau of Reclamation in the 1960s and early 1970s, imports water to the Rio Grande basin from the San Juan Basin for use in New Mexico. It is accounted separately from native Rio Grande water and provides water to help alleviate shortages in available native Rio Grande water. In the future San Juan–Chama water will be a primary source of drinking water to the citizens of Española, Los Alamos, Santa Fe, and Albuquerque.

The Bureau of Reclamation operates the project, diverting a portion of New Mexico's Upper Colorado River Compact apportionment from the San Juan Basin to the Rio Grande basin. The project provides additional surface water for New Mexico water users with San Juan–Chama contracts. The operation of the project requires complex accounting procedures. In order for the system to function properly, a partial adjudication of water rights was conducted on the Rio Chama by the Office of the State Engineer to protect San Juan–Chama Project water from being consumed by the acequias once it is released from upstream reservoirs. Additionally, the state engineer established bypass flow requirements through El Vado and Abiquiu Reservoirs to provide Rio Chama acequias their senior water rights downstream of Abiquiu Dam.

The San Juan–Chama Project brings added flexibility in managing water in the Upper Rio Grande basin above Elephant Butte Reservoir. It has been used to maintain the pool of water in Cochiti Reservoir, the recreation pool in Elephant Butte Reservoir, for irrigation, and to offset the effects on the river of pumping for municipal and industrial uses. The water has also been used to provide secondary benefits such as winter flows on the Rio Chama and to aid in meeting flow targets of the Endangered Species Act between its point of release and point of use.

THE 1950s DROUGHT AND U.S. SUPREME COURT LITIGATION

During the 1950s drought Texas sued New Mexico and New Mexico and Texas sued Colorado in the U.S. Supreme Court. The suits were filed to force New Mexico and Colorado to comply with the 1938 compact and make up under deliveries (then more than 300,000 acre-feet for New Mexico and 900,000 acre-feet for Colorado). The Texas case against New Mexico was dismissed on a technicality. Nonetheless, with one caveat, El Vado Reservoir has since been operated in compliance with the 1938 compact. The federal government, as part of its tribal trust responsibility to the six Middle Rio Grande pueblos, stores water in El Vado Reservoir as insurance for delivery of direct flow to the Prior and Paramount Lands (lands identified as having a senior water right to other Middle Rio Grande Conservancy District lands in the 1928 act of Congress) of the six Middle Rio Grande pueblos. The stored water is released for delivery when the direct flow of the river drops below levels the federal government has estimated to be needed to adequately deliver water to the Prior and Paramount Lands.

In 1968 the U.S. Supreme Court granted a stipulation for continuance of the New Mexico v. Colorado case as long as Colorado met its annual compact obligation until it was once again in compliance. Colorado met or exceeded its obligation each year from 1968 through 1984 and has remained in compliance since then. Its remaining under-delivery to New Mexico was cancelled in 1985 when Elephant Butte Reservoir spilled. The case was subsequently dismissed. To meet its annual obligation, Colorado restricts the diversion of surface water users with rights that pre-date the 1938 compact.

Suggested Reading

Upper Rio Grande Water Operations Review and Environmental Impact Statement. U.S. Bureau of Reclamation, U.S. Army Corps of Engineers, and New Mexico Interstate Stream Commission, 2005.

The Upper Rio Grande: A Guide to Decision-Making, Steven J. Shupe and John Folk-Williams, Western Network, Santa Fe, New Mexico, 1988.

The Administration of the Rio Grande Compact in Colorado, Steven E. Vandiver. CLE on the Law of the Rio Grande, 2003.

CHAPTER TWO

THE MIDDLE RIO GRANDE TODAY

DECISION-MAKERS FIELD CONFERENCE 2007
San Acacia to Elephant Butte

CHAPTER TWO

San Acacia Diversion Dam

Infrastructure and Management of the Middle Rio Grande

Leann Towne, *U.S. Bureau of Reclamation*

Many entities are involved in water management in the Middle Rio Grande valley from Cochiti to Elephant Butte Reservoir. These entities own and operate various infrastructure in the Middle Rio Grande valley that are highly interconnected and ultimately affect water management of the Rio Grande. This paper describes major hydrologic aspects of the Middle Rio Grande valley, including water management activities of the U.S. Bureau of Reclamation, major infrastructure of the Middle Rio Grande Project (including the Low Flow Conveyance Channel), and focusing on issues downstream of San Acacia Diversion Dam. Although other entities such as municipalities have significant water management responsibilities in the Middle Rio Grande valley, they will not be addressed in this paper.

The Middle Rio Grande Conservancy District, a political subdivision of the state of New Mexico, was formed in the 1920s as a conglomeration of a number of acequias (community irrigation ditches) and the six pueblos of the Middle Rio Grande valley. In 1950 Congress authorized the Middle Rio Grande Project to stabilize and improve the economy of the Middle Rio Grande valley by rehabilitation of the Middle Rio Grande District facilities and by controlling sedimentation and flooding on the Rio Grande. The U.S. Bureau of Reclamation and the Corps of Engineers jointly planned the development of the project. In the Middle Rio Grande valley, the Bureau of Reclamation undertook rehabilitation of project irrigation and drainage works and channel realignment, and the Corps of Engineers was responsible for flood protection.

MIDDLE RIO GRANDE VALLEY INFRASTRUCTURE

The Middle Rio Grande Project irrigation distribution and drainage system begins just below Cochiti Dam and extends to Elephant Butte Reservoir, 174 miles when the reservoir is at maximum capacity. The project is capable of supplying irrigation water for as many as 90,000 acres within the Middle Rio Grande valley including water for the Pueblos of Cochiti, Santo Domingo, San Felipe, Santa Ana, Sandia, and Isleta.

The Middle Rio Grande Conservancy District (MRGCD) consists of four divisions: Cochiti, Albuquerque, Belen, and Socorro, serving irrigated lands within the Middle Rio Grande valley from Cochiti Dam to the Bosque del Apache National Wildlife Refuge. The four divisions are served by Middle Rio Grande Project facilities, which consist of the floodway and three diversion dams, more than 780 miles of canals and laterals, and almost 400 miles of drains. Users are served by direct diversions from the Rio Grande and from internal project flows such as drain returns. These irrigation facilities are operated and maintained by MRGCD.

COCHITI DIVISION

Project diversions from the Rio Grande begin at Cochiti Dam, through two canal headings that serve the Cochiti Division. The Cochiti East Side Main and Sile Main canals deliver water to irrigators on both sides of the Rio Grande. Galisteo Creek and Tonque Arroyo, both east side tributaries, join the Rio Grande between Cochiti Dam and Angostura Diversion Dam. Diversions within the Cochiti Division primarily serve the Pueblos of Cochiti, Santo Domingo, San Felipe, and Santa Ana and the communities of Pena Blanca and Sile.

ALBUQUERQUE DIVISION

Angostura Diversion Dam, a concrete diversion dam, takes water from the Rio Grande to serve the Albuquerque Division of MRGCD. The Rio Grande from Angostura Diversion Dam to Isleta Pueblo is influenced by many factors including contributions from the Jemez River, which flows into the Rio Grande just downstream of Angostura Diversion Dam, municipalities adjacent to the river, and MRGCD operations. The city of Albuquerque waste water return flows and soon-to-be-completed drinking water diversion from the Rio Grande are within this reach. The Albuquerque Division of MRGCD serves the Pueblos of Sandia, Santa Ana, and Isleta, as well as non-Indian users in Bernalillo, Corrales, and Albuquerque.

BELEN DIVISION

The Belen Division of MRGCD diverts water from the Rio Grande at Isleta Diversion Dam, a concrete diver-

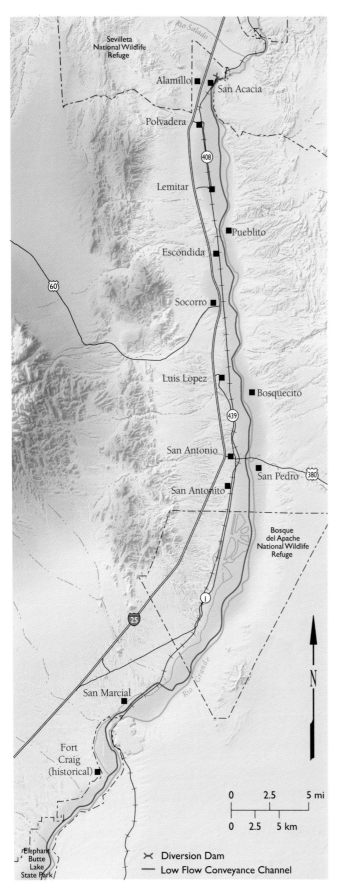

sion dam on Isleta Pueblo. The Rio Puerco and the Rio Salado enter the Rio Grande from the west side 42 and 51 miles respectively below Isleta Diversion Dam. These tributaries are a significant source of heavy sediment-laden flows to the Rio Grande, particularly from summer monsoon events and can be a major source of irrigation water to Socorro Division farmers when upstream supplies are limited. Delivery and drainage for the east side of the river ends in this reach where the Lower San Juan Riverside Drain outfalls to the river about three miles upstream of the confluence with the Rio Salado. The Belen Division serves Isleta Pueblo, several New Mexico Department of Game and Fish refuges, the Sevilleta National Wildlife Refuge, and irrigators in several communities including Bosque Farms, Los Lunas, Los Chaves, and Belen.

SOCORRO DIVISION

About 55 miles downstream of Isleta Diversion Dam, the final Rio Grande diversion for MRGCD is the San Acacia Diversion Dam, a concrete diversion dam that serves the Socorro Division. The Socorro Division also relies on flows from several sources within the project systems, including the Unit 7 Drain, supplied by surface water and ground water return flows from the Belen Division, and direct diversions from the Low Flow Conveyance Channel. The Socorro Division serves users in the Socorro area, and the U.S. Fish and Wildlife Service Bosque del Apache National Wildlife Refuge.

SAN ACACIA DIVERSION DAM TO ELEPHANT BUTTE RESERVOIR

As of October 2006 it was approximately 78 river miles from the San Acacia Diversion Dam to the Elephant Butte Reservoir pool. This distance varies depending on the elevation of Elephant Butte Reservoir. When Elephant Butte Reservoir is at maximum capacity, it is about 54 river miles from San Acacia Diversion Dam to the reservoir pool. There are no major tributaries to the Rio Grande below San Acacia Diversion Dam. However, several arroyos and two flood control channels enter the Rio Grande from both sides of the river. The east side of the river is constrained not by a levee or drain, but rather by the east mesa valley wall. The mainstream Rio Grande channel in this section of the river is roughly 200 to

Middle Rio Grande Project features.

600 feet wide and is meandering near San Acacia, becoming more braided downstream.

This stretch of river contains two parallel channels: the mainstem Rio Grande and the Low Flow Conveyance Channel. The Low Flow Conveyance Channel is located on the western side of the Rio Grande, runs parallel to it at a lower elevation, acting as a drain, and is protected from the river by a continuous spoil dike. As previously mentioned, there are diversions from the Low Flow Conveyance Channel for the MRGCD and the Bosque del Apache National Wildlife Refuge. There is a flow constriction approximately 45 miles downstream of San Acacia Diversion Dam near the San Marcial Railroad Bridge. This constriction is due to the river bed rising (aggradation) at the bridge. The Corps of Engineers is currently evaluating options for replacing the spoil dike protecting the Low Flow Conveyance Channel (LFCC) and relocation of the San Marcial Railroad Bridge.

Sediment deposition has a significant impact on this stretch of the river. High sediment loads in the Rio Grande have led to significant aggradation of the river from the Bosque del Apache National Wildlife Refuge to Elephant Butte Reservoir. Due to this aggradation, the river bed is generally above the level of the surrounding valley floor downstream of the refuge. This aggradation and other factors such as vegetation growth have made maintaining a connected river to the Elephant Butte Reservoir pool an issue since the early 1950s. River channelization projects have historically been needed to ensure delivery of water to Elephant Butte Reservoir. These projects are cooperatively funded and constructed between the Bureau of Reclamation and the state of New Mexico. At present more than 20 miles of temporary channel have been excavated since 2000. This temporary channel work has been crucial in maximizing delivery of water to Elephant Butte Reservoir. It is estimated that about 15,000 acre-feet of water per year has been saved because of the temporary channel work.

There is a complex hydrologic connection between the Rio Grande, the LFCC, and drainage from irrigation on the west side of the Middle Rio Grande valley. Middle Rio Grande Project facilities and drainage converge in this area west of the LFCC and eventually end up in the channel with one minor exception: irrigation drainage directly to the river. In addition, due to aggradation of the river, significant seepage takes place from the Rio Grande to the LFCC, which acts as a drain for the valley. Such a multifaceted hydrologic system leads to challenging operations that are described later.

LOW FLOW CONVEYANCE CHANNEL

The Middle Rio Grande Project's LFCC parallels the Rio Grande on the west side and originally extended from San Acacia Diversion Dam to the narrows of Elephant Butte Reservoir. The LFCC is owned, operated, and maintained by the Bureau of Reclamation. Construction began in 1951 and was completed in 1959. The LFCC was constructed to aid delivery of Rio Grande compact waters and sediment to Elephant Butte Reservoir. It also served to improve drainage and provide additional water for irrigation. It is a riprap-lined channel that parallels a 54-mile reach of the Rio Grande from San Acacia to San Marcial. The LFCC collects river seepage and irrigation surface and subsurface return flows, thus reducing evaporation. The usefulness of the LFCC is dependent upon the water level of Elephant Butte Reservoir. Depending upon the condition of the outfall, a maximum of 2,000 cfs can be diverted into the LFCC at San Acacia.

Diversions from the river into the LFCC began in 1953, and diversions at San Acacia began in 1960. With above average water years the reservoir was relatively full through the 1980s. During this time the lowest reaches of the LFCC, which were inundated by the reservoir, became filled with sediment. This made the outfall of the LFCC difficult to maintain, and therefore diversions ceased in 1985. Since that time the LFCC has carried only drainage and irrigation return flows, with minor exceptions.

Currently the spoil dike that protects the LFCC (and surrounding lands such as the Bosque del Apache National Wildlife Refuge) from Rio Grande flooding is threatened by overtopping downstream of the Bosque del Apache Wildlife Refuge because of sediment deposition in the river channel. Environmental groups have also raised concerns about the impacts of future LFCC operations on the bosque, wildlife resources, and endangered species in the river below San Acacia Diversion Dam. The states of Colorado, New Mexico, and Texas, and farmers in the lower Rio Grande have raised concerns that compact deliveries will be impaired if the LFCC is not operated. Due to these factors and the condition of the channel outlet, operations of the LFCC as originally intended are not currently possible.

In order to meet needs of the endangered Rio Grande silvery minnow, the Bureau of Reclamation began pumping from the LFCC into the Rio Grande at four locations in 2000. These pump sites begin approximately 20 miles downstream of San Acacia Diversion Dam at the Neil Cupp pump site. Two sites

> **A Few Definitions**
>
> *Canal*—a waterway channel designed for draining or irrigating land.
> *Diversion*—a structure constructed to divert water from one course or body of water to another.
> *Drain*—a canal designed for draining irrigated land.
> *Floodway*—an area which serves as a course of conveyance for the river.
> *Lateral*—a side canal designed for irrigating land.
> *Levee*—an embankment designed to prevent flooding.
> *Outfalls*—the outlet of a canal or drain to the river.
>   River recessions—the upstream receding of water within the river.
> *Spoil dike*—an embankment constructed, but not designed, to control and confine water.

are located at the northern and southern boundaries of the Bosque del Apache National Wildlife Refuge, approximately 6 and 16 miles downstream respectively from the Neil Cupp location. Finally pumping occurs at the Fort Craig site approximately 10 miles downstream from the southern boundary of the Bosque del Apache National Wildlife Refuge. Fifteen pumps are currently available to supplement Rio Grande flows and manage river recessions consistent with the current Biological Opinion.

ENDANGERED SPECIES OPERATIONS

The U.S. Fish and Wildlife Service issued a Biological Opinion on March 17, 2003, which affects water operations along the Rio Grande. The Biological Opinion contains flow targets for dry, average, and wet hydrologic years and when certain Rio Grande Compact restrictions affecting upstream storage are in place. The Biological Opinion focuses on keeping the Rio Grande continuously wet when possible between Cochiti Dam and Elephant Butte Reservoir to meet the needs of the Rio Grande silvery minnow. In average and dry years, and when certain Rio Grande Compact restrictions are in affect, the focus shifts to maintaining continuous flow to Elephant Butte Reservoir through the spawning period (June 15), with some flexibility to allow drying post-spawn.

The Bureau of Reclamation's Supplemental Water Program began in 1996 when supplemental water was first acquired and managed to augment river flows. Since then the program has been significantly expanded to assist in maintaining flows to meet Biological Opinion flow requirements. Two major components of the bureau's Supplemental Water Program include securing water from willing buyers/leasers to be released to the Rio Grande from upper-basin reservoirs, and pumping from the LFCC into the Rio Grande below San Acacia. The Supplemental Water Program supplies generally are released from storage in Heron, El Vado, or Abiquiu Reservoirs. It takes a minimum of seven days for water released from Heron and El Vado Reservoirs to get to San Acacia Diversion Dam and a minimum of five days for water released from Abiquiu Diversion Dam to reach San Acacia Diversion Dam.

Water management for endangered species from San Acacia to Elephant Butte Reservoir involves a multifaceted operation because of the complexity of the hydrology, the many entities involved, and ever-changing conditions. A key component impacting the hydrology and conditions in the area is losses in the system. Losses are defined as the reduction in quantity of water in transit not attributable to intended removal such as diversion. Losses in the Rio Grande are in large part a result of evaporation and consumption by riparian vegetation (evapotranspiration). These losses approach 50 percent of total depletions in the San Acacia reach and can vary greatly depending on weather conditions. The need for releases of supplemental water and pumping from the LFCC varies on a daily basis depending on many hydrologic conditions including losses in the system, diversions and return flows from water users, snowmelt runoff, and summer thunderstorm contributions to flow. With supplemental water supplies a minimum of five days travel time to San Acacia Diversion Dam, it can be difficult to maintain necessary endangered species flows. The LFCC is used to assist in maintaining those flows and to help manage river recession in times when continuous flows are not required under the Biological Opinion.

There will always be a need to manage the river above Elephant Butte Reservoir. The high sediment loads coupled with the relatively flat valley slope require regular channel maintenance to keep water flowing to the reservoir. With continuously changing river conditions, management of the river in this area will take careful evaluation and analysis.

The future of water management below San Acacia will likely evolve as decisions are made on water management needs, including endangered species needs, operations of the LFCC, and potential modifications of infrastructure such as the LFCC, associated levees, and the San Marcial Railroad Bridge. Such decisions will affect the hydrologic and environmental conditions through this reach; therefore, they will take careful consideration and evaluation and will require input from many entities.

Impacts of Middle Rio Grande Project Construction on the Middle Rio Grande Floodway—Rolf Schmidt-Petersen, *New Mexico Interstate Stream Commission*

The Middle Rio Grande Project was constructed and is operated to provide flood protection, control sediment movement, and improve valley drainage and downstream water deliveries within the valley from Velarde to Elephant Butte Reservoir. While the goals of the project were accomplished in part, the Rio Grande in the middle valley became less like a natural river system. The project also served to create and maintain the full canopy cottonwood bosque seen in parts of the valley today. Unfortunately and fortunately, it has had other effects, outlined below:

Cochiti Reach—Here the floodway is variable in width (up to 2,000 feet), bounded by discontinuous spoil bank levees and many small, relatively short riverside drains, which return water to the river at many points within the reach. The river flows in a narrow, deep channel up to several hundred feet wide, with few islands. Project infrastructure is used to deliver water to the pueblos of Cochiti, Santo Domingo, and San Felipe, and to farmers in Pena Blanca and Algodones. It protects agricultural lands, portions of the Cochiti, Santo Domingo and San Felipe Pueblos, and houses in the lower portions of the valley. Due partly to construction of project facilities, the river has incised within the originally constructed channel, reducing the potential for flooding. Due to the incision and downstream constraints on releases from Cochiti and Jemez Canyon Reservoirs, bosque flooding does not normally occur. Silvery minnow are present in small numbers.

Albuquerque Reach—The floodway here is more constrained (less than 1,500 feet wide) but the channel is somewhat wider (~600 feet), and many islands have formed over the past ten years of drought-induced lower flow. The floodway is bounded by spoil bank levees, a couple of engineered levees, and nearly continuous riverside drains. Project infrastructure is used to deliver water to the pueblos of Santa Ana, Sandia, and Isleta, and to farmers in Corrales and Albuquerque. River incision has reduced flooding potential by levee sloughing or overtopping, but the potential for flooding due to a bank erosion-caused levee breach remains. Because the bottom of the river is generally perched above lands outside the levees, a levee breach could have significant impacts. Flood protection issues have increased from the protection of agricultural lands and infrastructure and parts of downtown Albuquerque, to protecting people and their homes in much of the valley. Bosque flooding does not normally occur in this reach. This has prevented new cottonwood trees from germinating and reduced habitat for the silvery minnow. Balancing flood protection with ecosystem and endangered species preservation is an obvious challenge.

Isleta Reach—South of Albuquerque the floodway and river channel are approximately the same width as in the Albuquerque Division and are bounded by spoil bank levees and nearly continuous riverside drains. Project infrastructure is used to deliver water to the pueblo of Isleta, to farmers in the valley from Bosque Farms to Bernardo, and to the wildlife refuges at Bernardo and La Joya. Levee sloughing, overtopping and bank erosion of the levee are potential flood threats. Many islands have formed in the river over the past decade. Due to the shift from rural to suburban conditions in the valley, flood protection priorities in this reach are similar to those in the Albuquerque reach. Because the river channel has not incised significantly in this reach, significant areas of bosque can and are flooded under the maximum upstream flow releases. Thus, the potential for cottonwood reproduction is high, as is the potential for spawning and recruitment of the silvery minnow. In 2005 this reach of the river had some of the highest number of collected silvery minnow. However, flood control issues in the San Acacia reach constrain upstream flood control reservoir releases, which in turn limit the potential for additional flooding of the bosque in this reach and upstream.

San Acacia Reach—The San Acacia Division has a markedly different floodway configuration than the two reaches directly to the north. The river here is unconstrained by a levee on its east side. The floodway can be over 2,000 feet wide in places and the river channel quite variable in width (from 100 to over 1,000 feet). Several small discontinuous drains on the east side of the river serve to drain water from relatively small farmed areas back to the river. The LFCC currently serves as the riverside drain on the west side of the floodway. The LFCC is larger and deeper than most other riverside drains in the middle valley. South of Escondida, the LFCC does not return water to the river. Because of aggradation of the river bed, water in the LFCC is conveyed directly to Elephant Butte Reservoir. Significant bosque flooding can and does occur south of Escondida. Most irrigation, including that on the Bosque del Apache, occurs west of the floodway and is served by the Socorro Main Canal and the LFCC.

In sharp contrast to the reaches to the north, sediment is being deposited by the river, and the river bed has aggraded in the reach from just north of NM–380 south. In some places near San Marcial the bed of the river is 5–10 feet higher than the valley floor to the west and 2–3 feet higher than the valley floor to the west, creating a significant flood risk. Levee sloughing, overtopping, and bank erosion of the levee are potential flood threats. Significant amounts of money are spent each year by the Bureau of Reclamation and the ISC to keep the river channel open and reduce the risk of a levee failure. However, the existing flood risks significantly constrain upstream releases from the Corps of Engineers flood control reservoirs, which limits the potential for flooding of the bosque in upstream reaches.

The Middle Rio Grande Water Budget: A Debt Deferred

Deborah L. Hathaway and Karen MacClune, *S. S. Papadopulos & Associates*

Ground water has enabled increased water use over the past several decades without significant impact to the stream system; however, this practice is not without cost. As lagged impacts of past pumping reach the river, ever increasing offsets will be required. For a time, these offsets can be met with stored surface water, for example, Albuquerque's stored San Juan–Chama Project water in Abiquiu Reservoir. However, so long as water use exceeds water supply, balancing the water budget, again, is deferred to the future. Ultimately, a reduction in water use will be required to equal supply, or, alternatively, new supplies, not connected to the Rio Grande, must be acquired.

The Middle Rio Grande region (using a definition consistent with the Rio Grande Compact) extends along the Rio Grande from Cochiti Reservoir to Elephant Butte Reservoir, a distance of approximately 175 miles. Within this reach of the Rio Grande are many tributary streams and ground water basins. A water budget analysis for the region, including both surface water and ground water, indicates that, on average over the long term, the combination of regional water use and downstream compact obligation exceeds water inflow. At present, the difference is supplied by ground water stored in basin aquifers, as is reflected by declining ground water levels most notably in the Albuquerque reach. Regional water planners are projecting increased population, with commensurate increase in urban water use.

New Mexico pioneered the recognition of ground water-surface water interrelationships in water rights administration almost 50 years ago, when New Mexico State Engineer Steve Reynolds imposed controversial and unpopular requirements necessitating the offset of surface water impacts resulting from ground water pumping. This approach, in theory, would allow the utilization of what was then believed to be vast ground water resources, while keeping the river "whole." While painful to those who would prefer to pump ground water without considering surface water impacts, the system was clever. With this approach, significant additional development along the Rio Grande could and did occur, and compliance with the Rio Grande Compact, governing downstream delivery obligations, was manageable. State Engineer Reynolds understood that the plan would require ever-increasing offsets as stream depletion grew, and that, ultimately, the plan was not sustainable. However, his statutory charge obligated him to make water available to the public for beneficial use, so long as existing water rights were protected. Problems associated with growing stream depletions from ground water pumping were not imminent, and would not be problematic during planning periods (typically, 20 or 40 years) being addressed with water management actions at that time.

Today many stakeholders in the Middle Rio Grande understand that the existing pattern of water supply and water use is not sustainable. The lagged impact of ground water pumping on the stream system continues to grow. For various practical and political reasons, offsets don't always match the lagged impacts to the streams. And, it is recognized that the ground water resource itself has limits. At state, regional, and local levels, questions are asked: What is the water supply? What is the water demand? What are the consequences of making up deficits with ground water? And, what is the cost, economically, environmentally, socially, and culturally? Addressing these questions has been the focus of regional and state water planning over recent years. Support for this planning process has been the goal of the recent quantification of the Middle Rio Grande water budget by S.S. Papadopulos & Associates, Inc. in 2004.

THE WATER BUDGET: INFLOWS, OUTFLOWS, AND DEFICITS

The term "water budget" is commonly applied by hydrologists to mean an accounting of the inflow to, outflow from, and storage in a hydrologic unit such as a drainage basin or aquifer. In this fashion, the Middle Rio Grande water budget is characterized through examination of water supply from streams and ground water, and water use from both resources.

Surface water supply to the Middle Rio Grande region comes from several sources:

- Inflow from the Rio Grande main stem and native inflow from the Rio Chama, reflected in the gage at Otowi;

- Imported San Juan–Chama water, originating from the San Juan River basin in Colorado, also reflected at the Otowi gage; and,

- Inflow from tributaries to the Rio Grande between Otowi and Elephant Butte, most notably the Rio Jemez, Rio Puerco, and many monsoon-driven arroyos and rivers, including the Rio Salado.

The surface water supply varies dramatically from year to year. For example, between 1950 and 2002, the annual Rio Grande inflow at Otowi (excluding imported water from the San Juan basin, termed, San Juan–Chama inflow) ranged from 255,000 to 2,170,000 acre feet, with a mean value of 940,000 acre feet. Similarly, the tributary inflow is highly variable. The San Juan–Chama inflow, on the other hand, is relatively stable with about 81,000 acre-feet per year available to the Middle Rio Grande region, with Heron Reservoir used to store and regulate this supply

The ground water supply in the Middle Rio Grande is stored in basin aquifers through which the Rio Grande passes. Although the aquifers receive recharge along the mountains that bound the basins, the recharge is significantly smaller than the amount of ground water pumped. The aquifers function as underground storage reservoirs. As water is withdrawn, ground water levels drop, reflecting the removal of stored water; i.e., "mining" of ground water. Also occurring is some replacement of ground water from the surface water supply via stream depletion.

Water use in the Middle Rio Grande can be broken down into four main categories:

- Agricultural water use

- Municipal (including industrial and domestic use)

- Riparian water use (water use by non-cultivated vegetation along the river or other channels, in the bosque, and in receding reservoir pools)

- Open-water evaporative losses from rivers, ponds, and reservoirs

The pie chart of consumptive use in the Middle Rio Grande on this page shows the relative percentages of water use in these categories, including water use that is derived from both surface water and ground water. Of the total consumptive water use of about 700,000 acre-feet per year, the largest single use is riparian

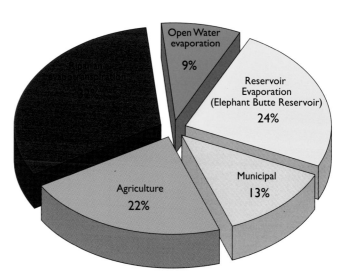

Consumptive water use on the Middle Rio Grande.

evapotranspiration (water used by plants in the riparian zone adjacent to the river) at 32 percent. Reservoir evaporation and agricultural consumptive use are estimated at 24 percent and 22 percent of the total, respectively. Consumptive urban use, including both surface water and ground water (minus wastewater returns to the river) represents 13 percent of water use. The category with greatest variability is the reservoir evaporation, which is dependent on the surface area of the reservoirs. Evaporation from Elephant Butte (including evapotranspiration from plants in drained areas when the reservoir level is low) has ranged from about 80,000 to 260,000 acre-feet per year over the 1950–2002 period.

A large percentage of inflow to the Middle Rio Grande is obligated to users below Elephant Butte Reservoir under terms of the Rio Grande Compact. The downstream obligation under this agreement is based on native inflow at the Otowi gage. For example, in rough terms, the delivery obligation is 57 percent of inflow at lower flows, i.e., below about 600,000 acre-feet per year; is about 60 to 62 percent of inflow at average inflows, i.e., in the range of 900,000 to 1,000,000 acre-feet per year; and is 80 percent when inflow is as high as 2,000,000 acre-feet per year. Using the schedule provided in the compact, one can calculate the average delivery obligation over a period of years, corresponding to any given set of inflows. For inflows such as those occurring between 1950 and 2002, an average of corresponding compact obligations is 645,000 acre-feet per year. Of course, this value varies dramatically from year to year.

CHAPTER TWO

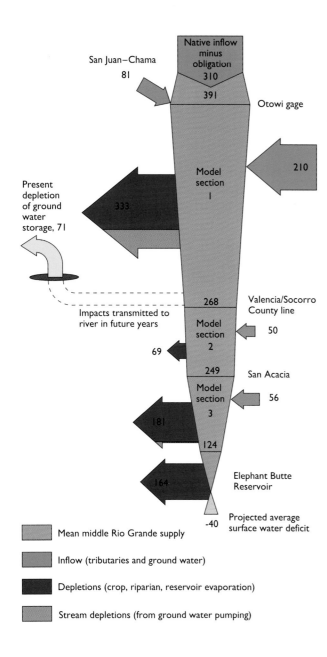

Water budget for the Middle Rio Grande, showing inflow and consumptive use along the stretch of river between Otowi gage and Elephant Butte Reservoir. Numbers are thousands of acre-feet.

A useful way to view the Middle Rio Grande water budget is to examine the portion of the supply that is available for consumptive use in the Middle Rio Grande region, apart from the compact delivery obligation. The water budget schematic on this page illustrates average inflow available to the region and consumptive uses along various reaches between Otowi and Elephant Butte, based on present levels of population and water use under a wide range of climate conditions, as developed from a modeling analysis conducted for the Middle Rio Grande Water Supply Study. In this schematic the compact obligation is subtracted from the upstream inflow at Otowi, to highlight the amount of flow available to New Mexico for consumptive use above the Elephant Butte Dam, including evaporation from the Elephant Butte Reservoir. Inflows and outflows are depicted, along with the "bottom line," a value representing surplus or deficit water at the downstream end of the region, or a likelihood of credit or debit under the compact. Using assumptions developed with the best available data and a wide range of potential climate conditions, on average, this region is projected to experience a shortfall of approximately 40,000 acre-feet per year in terms of surface water supply, and an additional deficit of 71,000 acre-feet per year occurs as a result of ground water pumping. This result does not represent specific conditions in any given year, nor does it necessarily translate to a compact violation. In reality, water management agencies work together with water users to avoid such a condition. With such efforts and other favorable conditions, despite the drought in recent years, New Mexico continues to maintain compact compliance. Nevertheless, and despite uncertainty in the water budget terms, the water budget modeling exercise underscores what has been the assumption by water management for decades: *The basin is fully appropriated.* New water uses impacting stream flows can only be supported by the cessation of old uses such that the overall consumptive use of stream flow does not increase.

BORROWING TO MAKE ENDS MEET: GROUND WATER

Presently, ground water in the amount of about 156,000 acre-feet per year is pumped throughout the Middle Rio Grande region, largely, for urban uses. Up to the present, to large degree, the stream impacts have been offset by wastewater inflows, thus, significant net depletion to surface flows has not occurred. However, the stream impact is lagged in time; the impact will grow over time, even if current pumping rates are held constant. The graph on the next page shows river depletions from historic pumping in the Middle Rio Grande valley, projected river depletions assuming current pumping rates are held constant into the future, and projected river depletions assuming implementation of the City of Albuquerque (Water Authority) Drinking Water Plan, whereby the Water Authority will reduce ground water pumping and supplement their supply with their San Juan–Chama water delivered via the Rio Grande. Under implementation of the Drinking Water Plan, which among other measures includes reduced usage per person, stream depletion from ground water pumping is reduced over about a twenty-year period, but then returns to an increasing trend. By the year 2040 stream depletions will have risen back to 2000 levels and will continue to increase.

Ground water has enabled increased water use over the past several decades without significant impact to the stream system; however, this practice is not without cost. As lagged impacts of past pumping reach the river, ever increasing offsets will be required. For a time, these offsets can be met with stored surface water, for example, Albuquerque's stored San Juan–Chama Project water in Abiquiu Reservoir. However, so long as water use exceeds water supply, balancing the water budget, again, is deferred to the future. Ultimately, a reduction in water use will be required to equal supply, or, alternatively, new supplies, not connected to the Rio Grande, must be acquired.

Alternatives

Exacerbating the balancing of the water budget is the fact that urban use is projected to increase due to population growth. Furthermore, some agricultural proponents are working toward reclaiming and irrigating additional lands. Many alternatives are being considered by the three planning regions within the Middle Rio Grande to address the anticipated growth. Some planning regions expect to retire agricultural land and transfer the water rights to urban use (and some planning regions are strongly opposed to this alternative). The reduction of riparian water uses within the bosque and reservoir delta is also proposed. In fact, the control of water depletion from riparian vegetation such as salt cedar has been an essential feature of water management for decades. For example, the construction and maintenance of agricultural drains, the Low Flow Conveyance Channel, and the Elephant Butte Pilot Channel have served to control water depletion from riparian vegetation. However, additional long-term depletions associated with Endangered Species Act requirements could make the riparian depletion control efforts more difficult and costly to perform. The ability to institute new riparian depletion control projects is likely limited. Several alternative water sources have been explored by Middle Rio Grande water stakeholders, including the use of desalinated water and cloud seeding. These

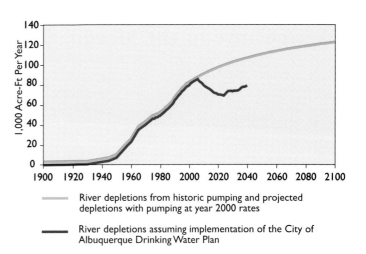

River depletions from historic pumping and projected depletions with pumping at year 2000 rates

River depletions assuming implementation of the City of Albuquerque Drinking Water Plan

options, presently in developmental stages, promise to be expensive.

Given that the amount of water consumptively used within the Middle Rio Grande is limited by the Rio Grande Compact, the surface water supply must be recognized as a singular and limited water supply. Significant work remains for planning regions to get onto "the same page" with respect to hydrologic, environmental, and cultural realities of alternatives for balancing the water budget. Implementation will be costly and will take time. However, many stakeholders and planners now recognize that the time has come to reconcile the water budget and effectively plan for the future. Forty-year planning horizons that were envisioned decades ago have come and gone.

Agriculture in the Middle Rio Grande Valley

Cecilia Rosacker-McCord, *Rio Grande Agricultural Land Trust*
James McCord, *Hydrosphere Resource Consultants*

The Rio Grande corridor in Socorro County contains the largest contiguous undeveloped tracts of farmland in the Middle Rio Grande valley. The river and adjacent farmland function as a linked hydrologic and ecologic system, providing habitat to the endangered silvery minnow and southwestern willow flycatcher and some of the most significant remaining cottonwood–willow forest or "bosque" in the Rio Grande basin (in fact in the entire southwestern U.S.). The farmland in this reach, together with the managed field crops and wetland habitat at Bosque del Apache National Wildlife Refuge, provides winter habitat to more than 100,000 migratory waterfowl of the Rio Grande flyway.

Farmland in the Middle Rio Grande valley is managed as small (less than 50 acres), medium (50 to 500 acres), and large (500 to 1,000 or more acres) farms. Socorro County operates more medium and large farms than the more populated counties of Valencia, Bernalillo, and Sandoval and cultivates more than 20,000 irrigated acres. The productive bottom lands of the Rio Grande produce some of New Mexico's most delicious green chile and melons, and most nutritious alfalfa hay.

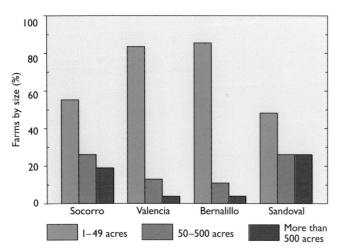

Farms (by size) in counties of the Middle Rio Grande valley. Socorro County is home to the greatest number of medium and large farms (more than 50 acres). Data from the U.S.D.A. 2002 Census of Agriculture.

Row crops provide an increasing percentage of cash crops in the Middle Rio Grande valley. Vegetables, melons, potatoes, and sweet potatoes account for $212,000 in sales in Socorro County alone.

The San Acacia reach stretches from the San Acacia Diversion Dam near the village of San Acacia southward to the Bosque del Apache National Wildlife Refuge and is contiguous with the Socorro Division of the Middle Rio Grande Conservancy District. In addition to describing the current status of agriculture in Socorro County, this article highlights ongoing trends and implications of those trends on the future of agriculture and local communities. The topics we focus on include acreage of farmland, loss of farmland to development, water rights and trends in water rights transfers, value of agricultural products grown in the area, and economic benefits of agriculture to communities of Socorro County. In addition, we identify some of the intangible benefits that agriculture provides to the area and the state and the potential public policies that may help preserve agriculture in the region.

IRRIGATED FARMLAND AND FARM ECONOMY IN SOCORRO COUNTY

The U.S. Department of Agriculture's 2002 Census of Agriculture, conducted by the National Agricultural Statistics Service, provides estimates of, among other things, the amount of irrigated farmland and farm income on a state and county basis. According to these data, Socorro County had an estimated 12,373 acres of irrigated land in 2002. In addition to this cropland, approximately 5,000 acres are irrigated at Bosque del Apache National Wildlife Refuge to grow feed crops and create moist-soil wetlands that benefit migratory waterfowl. The figure on this page summarizes irrigated acreage and market value of crops and livestock, both indicators of the health of irrigated farmland agriculture, and shows that irrigated farmland has been decreasing in Socorro County since 1992, but production value is steady or rising. In 2002 the total market value of crops and livestock in Socorro County, much of which is connected to farming in the valley of the Rio Grande, was $35,776,000, significantly more than any other county in the Middle Rio Grande valley.

Forage crops, primarily alfalfa, remain the number one cash crop in the Middle Rio Grande valley.

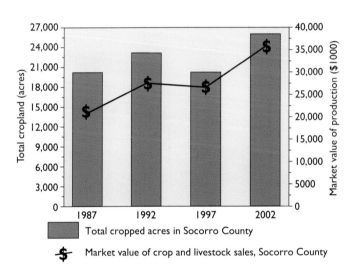

Historical trends in total irrigated crop acreage and farm production value in Socorro County. Data from the U.S.D.A. 2002 Census of Agriculture.

Socorro County produces a mix of crops. Of the 10,381 acres harvested in 2002, more than 9,000 acres (or about 90 percent) was cultivated in hay, grain, and grass forage, primarily alfalfa, which support the beef and dairy industry. Significant crops grown on the remaining acreage include feed corn, permanent pasture, and chile. In addition to livestock (cattle and horses) that graze on the permanent pasture, during the winter months many of the alfalfa fields are stocked with cattle trucked from nearby ranchlands. Since 2004 a significant increase in cattle prices has provided a healthy increase in the income for livestock and hay producers. Mixed vegetables grown for direct marketing (farmers' markets, farm stands, and local restaurants) typically occupy less than one percent of the total acreage, but they include an impressive and growing array of produce, including dry beans, snap beans, beets, carrots, eggplant, herbs, brassicas, potatoes, lettuce, melons, spinach, onions, peas, peppers, squash, sweet corn, tomatoes, and fruit. Chile and mixed vegetables have the potential to yield far greater net income ($2,000 to $10,000 per acre) than hay ($200 to $600 per acre), but require significantly more difficult-to-find farm labor. With the right equipment, one or two individuals can successfully operate literally hundreds of acres of hay land, whereas a vegetable grower requires a dozen or more laborers for 100 acres of chile.

HISTORICAL TRENDS IN WATER RIGHTS IN SOCORRO COUNTY

Surface water rights are generally associated with irrigated farmland, and that certainly holds for Socorro County. Other papers in this volume present information on water diversions, depletions, and water rights associated with irrigated lands. In this paper we highlight trends related to the transfer of water rights off of these lands and the impact that process has on the agricultural economy for the area. Because the waters of the Rio Grande are fully appropriated, increased demand by the growing population centers to the north of Socorro County, particularly the Albuquerque and Santa Fe metropolitan area, as well as the dramatic increase in the value of water rights (from $1,000/acre-foot in 1990 to more than $10,000/acre-foot today), a thriving market for transfer of water rights from the region has developed.

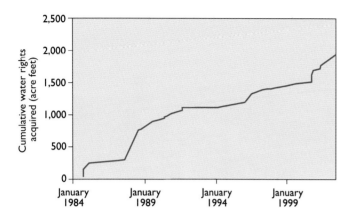

Time series graph of Socorro County irrigation water rights acquired by the city of Albuquerque (Office of the State Engineer file number RG-960).

Many such transfers have already occurred, and many applications are presently pending. For example, in the portion of the Socorro valley between San Acacia and Lemitar, we estimate that approximately 400 acres had their pre-1907 water rights sold and transferred before 2003, and more transfers certainly have occurred since then. Considering that this 7-mile reach of the valley has roughly 4,000 acres of farmland, approximately 10 percent of the farmland in Socorro's north valley has been sacrificed for the sake of growth in the Middle Rio Grande valley's metropolitan areas. One example of transfer of agricultural surface water rights to non-agricultural uses was obtained by reviewing the city of Albuquerque's water rights file with the Office of the State Engineer (file number RG-960). This file indicates that nearly 2,000 acre-feet of Socorro County irrigation water rights had been sold and transferred to the city of Albuquerque through the end of 2003. This represents more than 5 percent of the private irrigation rights in the county transferred to the city of Albuquerque alone through 2003.

Given that the Socorro–Sierra Regional Water Plan identified preservation of water for agriculture as one of its top three priorities, this trend should be worrisome to residents of the area. Evaluation of ways to minimize these transfers is an important component of the Socorro–Sierra Regional Water Plan, and local governments in the planning region are investigating their options for mitigating the adverse impacts of water rights exports. Obvious adverse impacts include the loss of irrigated farms and the many benefits they provide (see below), and the forsaking of all economic enterprises that depend on water (essentially everything), leaving the Socorro valley a dry and dusty, economically depressed area. This is similar to the situation faced by the communities in the Owens valley of California after the city of Los Angeles acquired their water.

AGRICULTURE PROVIDES INTANGIBLE BENEFITS TO LOCAL COMMUNITIES AND THE STATE

Visiting the farmland of Socorro County underlines the fact that this very narrow ribbon of arable land running alongside the Rio Grande (less than one mile wide in many places) is a limited and endangered resource for New Mexico. Historically we are an agrarian society, so New Mexico's original settlements lie along the river. As some of these original villages have grown into towns and cities, we now find that our agricultural lands are threatened due to their development potential and/or valuable appurtenant water rights. The families that have farmed these lands, their culture, and way of life are thus threatened and may

Agricultural lands provide critical wildlife habitat and open space in the Middle Rio Grande valley.

be lost forever. These lands offer the promise of a local, sustainable food supply in the future and are an incubator for a rich, uniquely New Mexican rural culture. The deterioration or collapse of rural communities results in negative social and cultural impacts not accounted for in traditional economic value analyses.

Apart from the fact that we grow the most delicious chile in the world, our farmlands produce a number of other products, including alfalfa, cotton, pecans, onions, lettuce, melons, apples, and just about any other fruit or vegetable you might find at the grocery stand. You need only visit our farmers' markets to experience the incredible variety of products grown here. Beyond their capacity for food production, these agricultural lands provide critical wildlife habitat and corridors, not to mention beautiful, scenic open space for all to enjoy. Vital rural communities rich in culture and traditions are essential ingredients for another of New Mexico's largest industries, as well: tourism. Tourism to the Bosque del Apache National Wildlife Refuge provides Socorro County one of its greatest sources of outside income.

Finally, consider that in this time of concern for homeland security it is now more critical than ever to keep our agricultural lands and food-producing capacity alive in all parts of the country, including New Mexico. While many Americans are rightly concerned about our energy security, most forget that as we lose our agricultural lands and the ability to feed ourselves, we become dependent on other countries for our food supply, at our own peril.

FUTURE TRENDS AND A POTENTIAL AGRICULTURAL PRESERVATION TOOL

Growing populations and dramatically increasing land and water rights values have created significant pressure to convert our productive lands out of agriculture. Many (if not most) of the agricultural landowners who have made a living off the land are approaching retirement age, and appreciated land and water values offer those landowners the promise of a comfortable retirement. As they approach their golden years, only two options currently exist for cash-strapped landowners: sell the land, or sell the water rights, to developers. Ultimately this means the loss of most of our irrigated farmland and the associated direct and intangible benefits described above.

Organizations such as Rio Grande Agricultural Land Trust are working with state and federal policy makers to develop a third option for landowners to realize the financial benefit of appreciated resource values, while at the same time preserving in perpetuity the agricultural, wildlife-habitat, and open-space values provided by the land. One of Rio Grande Agricultural Land Trust's efforts with policy makers is the development of a farmland preservation program for the state of New Mexico through what's known as a conservation easement. This is a non-development deed restriction that a landowner can elect to place on their land should they desire that it be maintained solely for agricultural production in the future. A landowner can donate a conservation easement on their land, and they will receive significant federal and state tax benefits. Alternatively they can sell the "development rights" to a qualified non-profit organization such as Rio Grande Agricultural Land Trust, and that organization would be obligated to monitor annually the easement and, if necessary, enforce it through the courts if a subsequent landowner attempted to violate the easement. Given that the value of development rights is typically appraised at about 70 percent of the full market value, purchasing them would take a fair amount of cash.

The U.S. Department of Agriculture's Farm and Ranchland Protection Program is a federal program that offers great promise for the farmers and ranchers of New Mexico. The Farm and Ranchland Protection Program provides compensation to landowners for their development rights and requires a minimum 50 percent cash match to the federal contribution. This type of "purchase of development rights" program for agricultural lands can be referred to as a Purchase of Agricultural Conservation Easement (or PACE) program. Rio Grande Agricultural Land Trust receives calls on a weekly basis from farmers, ranchers, and community leaders wanting to protect thousands of acres of agricultural land valued in the millions of dollars. Unfortunately, due to the lack of a state program in New Mexico that provides the requisite non-federal cash match, the federal program is almost non-existent here. Meanwhile the neighboring state of Colorado has received between $4 and $7 million of Farm and Ranchland Protection Program funds for each of the last several years, largely because of the state money dedicated for purchase of conservation easements that can be used for the required Farm and Ranchland Protection Program match. Rio Grande Agricultural Land Trust has thus been proactively engaging the governor's office and selected state legislators to develop a PACE program for New Mexico.

The Middle Rio Grande Ecosystem through the San Acacia Reach

Lisa M. Ellis, *Department of Biology, University of New Mexico*

In an arid region where water is limited, the Rio Grande once created a patchwork of floodplain plant communities that provided habitats for a rich diversity of animals. Essential to sustaining the diversity was the dynamic ebb and flow of the river, characterized by high spring floods alternating with periods of low discharge. Historically, such seasonal changes in river discharge orchestrated key ecological processes within the Middle Rio Grande ecosystem's floodplain and river. However, floodplain habitats along the Middle Rio Grande have decreased in recent years, largely because the hydrologic connection between the river and its floodplain has been disrupted by various forms of human activity. Despite the great decrease in habitat diversity, the San Acacia reach (between the San Acacia Diversion Dam and Elephant Butte Reservoir) retains a high potential for restoration.

RIVER CONNECTIVITY AND THE FLOOD PULSE

After more than a century of trying to modify and control the flow of large floodplain rivers, we now understand the importance of seasonal changes in water levels and how these help to maintain a river ecosystem's biological function and diversity. Along large rivers such as the Middle Rio Grande, periodic soil saturation by overbank or seep flooding—the "flood pulse"—regulates such functions as riparian (riverside) production, decomposition, consumption, and succession, all processes central to the ecosystem's integrity. In the Middle Rio Grande valley, as in other parts of the world, there is now an effort being made to restore the connectivity between rivers and their floodplains, with a particular emphasis on reestablishing natural flow regimes that drive the functioning of the rivers' riparian communities.

Thus along the Middle Rio Grande, including the San Acacia reach, reestablishing the natural pulse of the river is essential to restoring ecosystem processes. Before extensive river alteration, this reach was characterized by a broad floodplain supporting an ever-changing mosaic of habitats including patches of cottonwoods and willows of varying ages, open grassy meadows, marshes, ponds, sparsely vegetated akali flats, and open sandbars. This habitat mosaic was maintained by the considerable variation in river flow. The essential feature of the natural hydrograph for the Middle Rio Grande was a spring flush of water, which typically occurred in late May as a result of spring snowmelt from surrounding mountains. In some years this would have been a gentle rising of water to inundate forests, meadows, and other low-lying areas, but in other years high discharge meant scouring floods that cleared vegetation off sandbars and created optimal sites for cottonwood and willow regeneration. These scouring floods also transported sediment and reworked the channel, thus changing the appearance and position of the river bars. Open, moist sandbars are required for cottonwood and willow regeneration, because young seedlings cannot grow under the shade of a mature forest. Considerable sediment enters the Rio Grande via the Rio Puerco and Rio Salado. The San Acacia reach, lying below these confluences, benefits from this sediment.

The low-gradient, broad floodplain once favored a meandering channel, which helped to create the patchwork of habitat types. For example, old loops of the river were cut off to form oxbow ponds and eventually marshes and wet meadows. The connection between the river and ground water was and still is important for maintaining these habitats, because it wets the otherwise unsaturated soil above the water table.

Within the forest, the inundation of the rising river promotes a variety of ecosystem functions that are essential to maintaining the various floodplain habitats. For example, flooding increases rates of wood and leaf decomposition. In sites that are flooded regularly, increased rates of decomposition and subsequent mineralization (brought about by promoting fungal and bacterial activity) mean that nutrients are cycled back into the system more quickly, promoting new plant growth. As floodwaters percolate through the forest, impurities are filtered out and water quality increased. The deposition of sediments within the forest also provides an influx of beneficial nutrients.

Periods of low flow are also important to the system. Low water levels help to maintain suitable water temperatures, dissolved oxygen, and other water chemistry levels. Drawdown rates following flooding determine the fate of young plants beginning to germinate on the still-moist but exposed soils. Low flows also enable fish to move between feeding and spawning grounds and allow fish and amphibian eggs to remain suspended in the slow-flowing water column.

AN ECOSYSTEM DISRUPTED—LOSS OF THE FLOOD PULSE

We now know that high demands on river flow for irrigation withdrawals, along with efforts to constrain and straighten the river channel, impoundment of water, and even overgrazing in the upper watershed, have contributed to ecosystem degradation along the Middle Rio Grande valley. Along the San Acacia reach this degradation has resulted in significant loss of floodplain habitats. Reduction in peak discharge, along with early direct efforts to straighten the river, has greatly changed channel morphology. Over time, the channel has become narrower, more incised, and less active. New sandbar creation has decreased, and older bars are no longer scoured by high spring floods. This means that germination sites for cottonwoods and willows have greatly decreased, and remaining sites are commonly within the active channel, where young trees are killed by rising water. Without adequate germination sites, the forest cannot regenerate sufficiently to maintain itself, and as old trees senesce, they will not be replaced. Eventually, the cottonwood bosque will be lost.

The absence of flooding is also noticeable within the forest. Lack of seasonal inundation has greatly slowed leaf and wood decomposition, leaving increasing piles of organic debris and removing valuable nutrients from the system for long periods of time. Meanwhile, fungal and microbial activity slows down under dry conditions. The absence of floodwaters means that nutrient-laden sediments are not brought into the forest. Mature cottonwoods and willows are very sensitive to drops in the water table. Because they are phreatophytic, meaning their roots reach down into ground water, these trees are particularly affected by desiccating conditions resulting from water regulation.

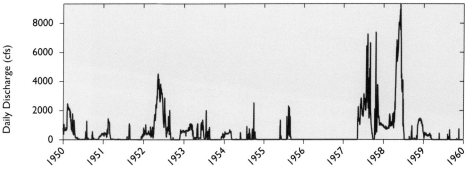

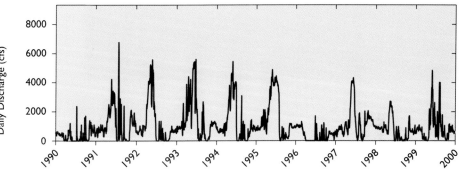

Daily discharge for the Rio Grande Floodway at San Marcial from 1950–59 (top) and from 1990–99 (bottom). Although fairly extensive regulation was already in place during the first decade, the upper graph still illustrates greater variability in flow, with higher highs and lower lows compared to the second decade, which was recorded after Cochiti Dam was built.

Cottonwoods are known to have reduced growth or increased mortality below dams or diversions, and to be especially susceptible to branch die-back induced by drought. Thus in addition to the decrease in regeneration seen with changes in sandbar morphology, increased mortality of adult cottonwoods has greatly diminished the quality of riparian habitats.

Further, the buildup of woody debris, which results from the increased branch death of cottonwoods during drought combined with decreased decomposition rates, promotes the spread of wildfires. Fire severity can be quite high in cottonwood forests with large amounts of dead and downed woody debris, and cottonwood survival in such fires is quite low. The loss of wetlands and wet meadows, together with their replacement by mainly exotic (non-native) trees and a more continuous forest constrained within levees and bluffs, means that natural fire breaks once inherent in the system are no longer in place. Fires also are fueled by the abundance of these exotics, particularly salt cedar.

Salt cedar (or tamarisk) is a particular problem along the San Acacia reach. Large sections of the floodplain here are covered only in salt cedar, to the

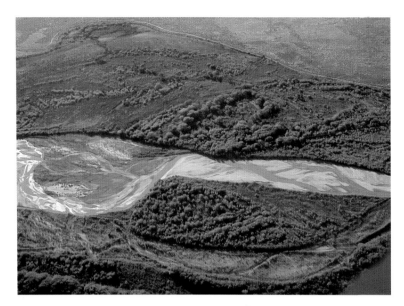

Aerial view of the Rio Grande in the Bosque del Apache National Wildlife Refuge, showing mature cottonwood bosque, wetland habitats, and some channel braiding. Note the old channel in the bottom of the picture. Sites like this have high restoration potential through restored flooding.

exclusion of other plants. Salt cedar can easily colonize areas disrupted by fire or other means, and in doing so, out-compete native cottonwoods and willows. Salt cedar is generally considered to be low-quality habitat for birds, and where it forms extensive monotypic stands (having only one type of plant), bird diversity is reduced. Much of the reduced quality of terrestrial habitats along the San Acacia reach is due to the presence of salt cedar. It affects aquatic habitats by stabilizing the riverbank and trapping sediments, which raises the level of the bank and reduces the potential for overbank flooding.

HABITAT DIVERSITY SUPPORTS WILDLIFE DIVERSITY

Riparian and floodplain habitats in arid regions are particularly valuable to wildlife. These habitats provide scarce resources not available in surrounding uplands, such as water, soils that allow digging, unique plant community types (e.g., large trees), other animals for prey, and more favorable ambient temperatures. They support a particularly high diversity of animal species, and often high population sizes. For example, at least 14 amphibian species, 29 reptile species, more than 200 bird species, and more than 50 mammal species have been observed within the Rio Grande floodplain of Socorro County. Invertebrate life is particularly well represented. At the Bosque del Apache National Wildlife Refuge at least 187 taxa of surface-active arthropods (insects, spiders, and their relatives) were identified in cottonwood sites alone.

The bird diversity of the San Acacia reach is quite rich, partly because the mosaic of plant communities provides opportunities for a variety of types of birds. For example, mature cottonwoods support species such as timber gleaning nuthatches and cavity-nesting woodpeckers and chickadees. Open sites with sunflowers or other non-woody plants are favored by seed-eating finches and sparrows, whereas sandbars and mud flats support large numbers of shorebirds. Marshes contain a number of wetland specialists, including blackbirds, herons, and rails. Ponds are filled with a variety of ducks, and cultivated fields are favored by flocks of geese and cranes. The importance of each habitat varies somewhat seasonally, depending upon breeding, migration, and over-wintering behavior. Neotropical migrant land birds breed here during the spring and summer, migrating shorebirds stop in large numbers during the spring and fall, and waterfowl are abundant during the winter. Taken together, the regional diversity of birds is extremely high due to the presence of all of these floodplain habitats, but this is threatened by the loss of habitat diversity. For example, the spread of salt cedar across the floodplain, to form monotypic stands that replace native habitats, decreases the options available to birds and other animals.

The San Acacia reach of the Rio Grande provides warm-water fish habitat, but channel narrowing and vegetation encroachment have decreased its favorable habitat characteristics. The reach supports at least 16 fish species, including natives and non-natives, but overall diversity has decreased due to extinctions and extirpations. A few species that do well during low-flow conditions, or recolonize rapidly after river drying, now dominate.

The loss of floodplain habitats has impacts on all wildlife in the San Acacia reach, but implications for threatened and endangered species are particularly great. Two federally listed endangered species have received much attention in recent years, the southwestern willow flycatcher and the Rio Grande silvery minnow. Both species have suffered greatly due to habitat loss, and both would benefit by restoration of the flood pulse. These two cases are prime examples of species needing ecosystem protection and illustrate the problems faced by all species. The southwestern willow flycatcher breeds in habitats with dense riparian vegetation, typically willows, seepwillows, or other shrubs or medium-sized trees, and very near to standing water or saturated soils. Overbank flooding is especially important in maintaining this vegetation and is one of the main factors determining habitat suitability. Restoring the flood pulse will increase habitat for this species.

The Rio Grande silvery minnow was once widespread along the Rio Grande south of Española, but now is restricted to the Angostura (Albuquerque), Isleta, and San Acacia reaches of the Middle Rio Grande. Until recent channel drying, the San Acacia reach supported the largest population and has been included in designated critical habitat. During much of the year, silvery minnows prefer pools, backwater, and secondary channels, habitats that are now limited in abundance. There appears to be a strong, positive correlation between silvery minnow abundance and river discharge. Silvery minnows spawn in response to the spring/early summer spike of water discharge—another direct link to the flood pulse. During years of low river flow, the river channel along the San Acacia reach often dries up completely, thus eliminating aquatic habitats. Extensive salvage and captive breeding efforts have saved the silvery minnow to date; however, the reconnection of the river to the floodplain is needed for its ongoing survival.

The San Acacia reach retains a significant amount of biological diversity and has definite potential for restoration. Although vulnerable to channel drying during years of low river discharge, it also experiences flooding at moderate flows. The northern end of the reach is incised and requires approximately 10,000 cfs discharge to initiate significant flooding, but the middle and lower portions of the San Acacia reach experience flooding at 3,000 cfs and 2,700 cfs, respectfully. The area of Bosque del Apache National Wildlife Refuge is particularly favorable to flooding and retains some floodplain connectivity. In some areas, a wide channel remains. The reach benefits from limited human development within the floodplain. Removing salt cedar, enhancing channel dynamics, and restoring flooding are all feasible and will contribute greatly to maintaining the high biological diversity of the region.

The dry river bed in the Bosque del Apache National Wildlife Refuge during the summer of 2005. Extensive channel drying in the San Acacia reach puts the endangered Rio Grande silvery minnow, as well as other aquatic species, in great peril.

The author wishes to thank Dr. Cliff Crawford and Gina Dello Russo for comments on this manuscript, Dr. Tom Turner for an update on the Rio Grande silvery minnow, and Dr. Chris Young for making the hydrograph.

The Middle Rio Grande—Short on Water, Long on Legal Uncertainties

G. Emlen Hall, *School of Law, University of New Mexico*

If you go out and look at the Middle Rio Grande these days, you'll see one stream of water flowing down the river. However, this river basin is one of the longest-settled, deeply hydraulic areas in the world. The river that runs through it reflects the many layers of governance that have overlaid the river in the more than 500 years that the river has maintained itself and supported human settlements dependent on it. In its long history of human use, the river has supported and been governed by the laws of Native American sovereigns, the Spanish Crown, the Mexican Republic, the territory of New Mexico, the state of New Mexico, and the United States. The laws of these different sovereigns have been laid down one on top of another and are often poorly integrated, if integrated at all.

These days we are accustomed to viewing the Middle Rio Grande surface flows as being made up of several different kinds of water: native flow, flood flow, stored flow, imported flow, and even flow contributed by interconnected ground water flows. These different flows reflect different sources for the single body of water we see when we stand on the Central Avenue bridge and watch the water come down. Imagine how much more complex the layered situation is when we consider that these different flows are themselves subject to different legal claims: Pueblo claims under Native American law, ancient acequia claims under Spanish and Mexican law, Middle Rio Grande Conservancy District claims under state law, Endangered Species Act claims under federal law.

If we're New Mexicans, bred to the basic law of the state, we tend to think that the law of prior appropriation governs these different sources of water established at different times. That law of prior appropriation said that rights to water were established by capturing water and applying it to beneficial use. The earlier you captured the water, the better your right to the common source shared with many others. Over the years we've tried to shoehorn all manner of different claims into the prior appropriation slipper. These days we talk about "non-consumptive beneficial uses," even though twenty years ago that term would have struck some listeners as an oxymoron. These days we acknowledge that restoration of rivers like the Rio Grande to more natural conditions is important, but we struggle to make sure a more natural flow does not increase depletions in the stream system, as the prior appropriation system says it should not.

Despite these and other Herculean efforts to cabin and contain ancient claims and new claims within the doctrine of prior appropriation, the boundaries are breaking down everywhere: Base flow may be subject to priority administration, but stored federal water is not; Pueblo water rights may be subject to federal control, but certainly are not governed by state law. Imported San Juan–Chama water may lie outside both state and Pueblo law. We don't know the answers to these puzzles.

In the end it's what we don't know about the simple and scant waters of the Rio Grande that may be the most important thing. How many basic facts about legal claims to the Middle Rio Grande don't we know at a time when the river is becoming more and more crucial to life here? Let me count a few.

PUEBLO WATER RIGHTS ON THE RIO GRANDE

Start with the six Middle Rio Grande pueblos. This is as good a place as any to begin, because the Pueblo claims to the river are so basic, so unknown, so unquantified, and so potentially large. We are accustomed to say that the 1938 Rio Grande Compact limits New Mexico's access to surface and ground water in the Rio Grande generally and in the Middle Rio Grande in particular. I'll come to the compact next, but the Pueblo claims come before the compact, which exempts them from its terms.

No one has ever formally and finally determined the nature and extent of Pueblo water rights outside the Middle Rio Grande although the numbers are everywhere. The Abeyta and Aamodt proposed settlements from Taos and Nambe don't tell us much, although, like most settlements, they vary from the formal law a lot. For example, under the proposed Aamodt settlement, the San Ildefonso Pueblo would agree to a senior surface water priority of a little less than 80 acres. Judge Mechem's 1985–1987 district court rulings said that the pueblos of Nambe, Pojoaque, and Tesuque had a prior and paramount right to their historically irrigated acres as of June 6, 1924, and quantified those. The San Ildefonso Pueblo would have received 365 acres of prior and paramount rights to the surface flows of the stream system and the interrelated ground water. This unique decision, establishing what the lawyers call the Mechem Doctrine, was never appealed and never confirmed by any higher court. Mechem based the general Pueblo doctrine he invented on the 1928 act by which Congress authorized inclusion of the six Middle Rio Grande pueblos in the Middle Rio Grande Conservancy District (MRGCD). That act confirmed the prior and paramount rights of those pueblos to some 8,800 acres (and a right equal with other district lands to an additional 12,000 acres), and Mechem found that the 8,800 acres represented the best evidence of the 1924 historically irrigated acreage of those pueblos.

The pueblos have always maintained that this congressional limitation on their prior and paramount rights applies only to water delivered through the MRGCD works. Other water, including ground water and water diverted from works other than the MRGCD's, was not limited by the 1928 act. At least three Tenth Circuit Court of Appeals judges who considered the nature and extent of the Pueblo water rights in 1976 thought that they probably had Winter's Rights, an expansive federal right based on practicably irrigable acreage, not actual irrigation.

Who is right in this tangled 75-year complication of half steps and missteps and stumbles and paralysis? No one yet knows. The range of possibilities still runs from the minimal rights accorded under state law to the maximum allowed under the Homeland and Practicably Irrigable Acreage standards of the full-scale Winter's Doctrine. If state law is the rule (and it seems highly unlikely), then the rights of the Pueblos would be minimal and manageable. If Winter's doctrine is applied, the Isleta Pueblo alone would command the whole flow of the Middle Rio Grande and its related ground water. If the Pueblos exercised their maximum rights to deplete the river, those new depletions would probably be charged to New Mexico under the 1938 Rio Grande Compact. New Mexico could not possibly meet its compact delivery obligations without catastrophically reducing other net depletions. That's a wide range of unknowns, and it's entirely outside what we call the primary limits on ground water imposed by the 1938 compact.

THE 1938 RIO GRANDE COMPACT

For the last decade we've paid lip service to the fact that the 1938 Rio Grande Compact obligations fundamentally limit New Mexico's net depletion of Rio Grande waters. The 1938 Hinderlider case teaches us that compact obligations come before any state law rights come into play. The 1938 compact shows us that New Mexico's obligations are measured by an index inflow at Otowi and a computed outflow below Elephant Butte. The gage relationships are based on 1929 conditions on the river. New Mexico is responsible for everything that happens between Otowi and Elephant Butte—everything being direct agricultural depletions, indirect municipal depletions, natural depletions, riparian depletions, and Pueblo Indian depletions. We used to think that winter snowpack above Otowi or summer torrential rains below Elephant Butte controlled New Mexico's widely varying deliveries and are allowable under the compact. But in the last 40 years, claims on the river have changed: Agriculture is down, municipal obligations to the river are up, and environmental attention to the river is producing its own set of new claims. The compact is blind to the causes of those increases and decreases. Based on flows as it is, the compact only sees allowable net depletions and sets absolute limits on the balance of those that control river flows. The sum of net depletions can't change, but the balance of them certainly can and has. As I noted before, New Mexico is responsible for all net depletions and is especially responsible for shifts and increases in net depletions created by rules of law. And these rules we don't know much about.

MUNICIPAL WATER

New Mexico fed the astronomical growth of its principal cities from 1950 to today primarily on stored ground water. New Mexico led the rest of the western states in its recognition that ground water is related to and responsible for interrelated surface water flows. The basic rule was this: Because depletions of surface water flows were limited to those allowed by the compact, New Mexico could switch uses from agriculture to municipal use, for example, but net depletions couldn't increase for any reason between Otowi and Elephant Butte. For a while existing agricultural uses, increasing municipal demand, and even growing riparian consumption could coexist, but eventually they would have to be re-ordered.

That eventuality has arrived. New Mexico knows that it has to react but hasn't decided how. Two recent examples from our biggest municipalities show the quandary we are now in. The 2000 Rio Rancho permit, which allowed the city to increase its ground water diversions to 25,000 acre-feet a year, put the city in hock to the Rio Grande for that full amount.

Initially the Office of the State Engineer tried to call the whole loan at present and at once, telling Rio Rancho that it couldn't pump any of the increase it had won until it offset all the future effects on the river. Eventually the state and the city agreed on a less draconian, staged series of offsets. The fact remains that Rio Rancho must reduce depletions on the Rio Grande by 25,000 acre-feet in order to leave the river in its compact-defined whole. That equals 12,000 acres of irrigated land, whole reaches of bosque riparian vegetation, or some other source. But it's got to be wet water. Nothing else will satisfy those compact gages.

We don't now know where the water will come from. I chatted once with an old time, retired, astute northern New Mexican who went door to door in Tesuque looking to buy a mere 5 acre-feet of wet water rights there at any price. He personally knew most of the residents. No one offered to sell. Some old time friends shut the door in his face, so offended were they at even the request that they sell their water rights.

There doesn't seem to be a much bigger voluntary market for old, secure water rights in the Rio Grande valley. You've heard the debates and proposed legislative limitations from the New Mexico governor and legislators on municipal power of eminent domain over land. Imagine the hell there would be to pay if Rio Rancho went into Socorro County to force by condemnation the transfer to the city of ancient surface water rights of farmers there. I once asked the Rio Rancho water attorney what she thought would give. "Something will change," she said. But she didn't know what, and neither do I.

In some ways the city of Albuquerque's recent fundamental switch from stored ground water to surface water represents the opposite side of the municipal coin. Deciding that its reliance on mining stored ground water was not sustainable, the city will switch to pumping a cocktail of imported San Juan–Chama stored water and ground water to satisfy a growing urban demand. San Juan–Chama water doesn't count in fixing New Mexico's delivery obligations at Otowi under the compact, but that water does show up at Elephant Butte and can help in meeting the state's obligation there. Switching to San Juan–Chama surface water will reduce that compact gift. At the same time, the switch will reduce another incidental contribution to New Mexico's compact obligation, the stored ground water that the city has pumped for 50 years, run through its municipal system, and discharged to the Rio Grande as a gift to the state's compact flows. Sandia National Laboratory computer models suggest that the combined loss of San Juan–Chama water and stored ground water may throw New Mexico off as much as 50,000 acre-feet per year in compact deliveries. But you don't need to go that far to recognize that municipal demand is already altering the balance of inputs and outputs in the middle reach of the river in fundamental ways and making deep commitments to the future of the river. We just don't know how much, and we don't know where the new balances leave us with respect to our fundamental compact obligations. In considering the city's newest application, the Office of the State Engineer's hearing examiner didn't say much about this rebalancing and said nothing real about the effect of the change on New Mexico's fundamental compact obligations.

THE MIDDLE RIO GRANDE CONSERVANCY DISTRICT

In the streets you often hear suggestions that it will require regional governance to solve the Middle Rio Grande problems, governance to match the width and depth of the resource problem. But for 75 years we've had a regional water government, the MRGCD. Thus far the MRGCD has contributed more to the problem of unknown claims to the river than it has contributed to a regional solution to the shifting demands, although the MRGCD has moved positively in this direction in the last decade. In recent public pronouncements the MRGCD has claimed a right to the water needed to irrigate 123,000 acres in the Middle Rio Grande. The authority for that number used by the district is thin indeed. It's not clear whether the claimed right is for diversion alone or for the full beneficial use of the water needed to irrigate the 123,000 acres. The fact that the MRGCD's recent statistics show that only about half of those acres, 61,000, are currently irrigated through district works would give anyone pause. Does the MRGCD "own" the water rights needed to irrigate the present acreage? Does the MRGCD "own" the water needed to irrigate the other dry 62,000 acres? Can the MRGCD and its water bank control that much water?

Two basic problems make it hard to see the MRGCD clearly. The first is the basic historical situation on which the MRGCD was overlaid in 1924. The best historical estimates suggest that irrigation in the Middle Rio Grande reached its peak in 1880 when approximately 120,000 acres were irrigated. A familiar set of hydrologic problems—poor drainage, an aggraded river bed, a diminished supply—contributed to a sharp drop in irrigation between 1880 and 1928 when the new district issued its plan for improvements. The MRGCD's vast and effective system of diversions and drains did bring back a lot of these acres, but to whom? The district, or the land owners of appurtenant water rights, or some hybrid of the two?

The statutory charter of the MRGCD compounded the basic confusion. The 1914 Ohio statute on which New Mexico's 1923 and 1927 statutes were based said that any improvements made by the new Ohio conservancy districts belonged to the districts. The statutory corollary in the Ohio statute said that landowners within the new districts could only claim ownership of uses that they could have made without the benefit of the conservancy district statute. These Ohio provisions made their way awkwardly into the 1923 and 1927 New Mexico statutes. They were subject to supplemental amendments that are too complicated to detail here. Suffice it to say that the MRGCD Magna Carta is hardly clear about the nature and extent of MRGCD ownership of and control over Rio Grande water.

FEDERAL CLAIMS TO RIO GRANDE WATER

Everybody knows about the tizzy into which the silvery minnow and the Endangered Species Act have thrown Rio Grande water managers. Thus far, a series of lawsuits, congressional acts, state law permits, and federal dam operations have been cobbled together to yield a tenuous compromise. The minnow may just be the beginning of federal alterations to inputs and outputs, to incidental contributions, and to unintended depletions to Middle Rio Grande surface flows. What remains unclear is how new federal operations will impact the 405,000 acre-foot maximum depletion that is allowed the Middle Rio Grande under the 1938 compact. Existing old compacts like the 1938 Rio Grande Compact don't even mention new federal claims. Suggestions for new compacts, none of which have yet been adopted, universally recommend making the federal government a more formal party to the agreements than it has been in the past, defining the federal rights and apportioning the effects of them among compacting states. Without such formal recognition, the federal government remains free under the supremacy clause and its own federal mandates to rework river flows. Which state pays for the federal changes if they effect state compact obligations, as they surely will?

We're accustomed in these post-silvery minnow days to look upstream on the Rio Grande, to look to the San Juan–Chama Project imported waters, or to look to federal dams at El Vado, Heron, and Abiquiu as the source of boons or bains to Rio Grande flows, depending on your perspective. The upstream problem is so potentially significant that we even have academic conferences on dam operations on the Rio Grande and its tributaries. But let me end by looking downstream at the effect of federal operations not so much on upstream inflows as on downstream deliveries.

Of course, I'm referring to the controversial Low Flow Conveyance Channel just above Elephant Butte. Put in almost 60 years ago by the Bureau of Reclamation as a way of increasing New Mexico's compact deliveries at a time when the state was falling farther and farther behind, the Low Flow Conveyance Channel seemed to work. New Mexico made up for its compact under-deliveries with the help of the channel. However, one of the unintended consequences of the Low Flow Conveyance Channel was a drying of the river itself in the reach below the Low Flow Conveyance Channel's intake. Long a legitimate target of environmental complaints about the harm done to the river itself by man-made alterations like the channel, the Bureau of Reclamation, especially since the turn of this century, has used the channel less and less. And one of the unintended consequences of that return to natural flows may well turn out to be compact under-deliveries for which the state of New Mexico will be responsible.

A SHORT END TO A LONG AND NEVER-ENDING STORY

The sum of all of these uncertainties—the nature and extent of pueblo and MRGCD rights, the source of rights for increasing municipal demand, the unintended consequences of changes to policies—is even greater uncertainty. On the wall of the office of the deputy chief counsel of the Office of the State Engineer is a map showing the surface water districts in New Mexico. It looks like a donut. Districts to the south, west, north, and east of the Rio Grande surround a hole in the center: the Middle Rio Grande from Otowi to Elephant Butte. From this perspective, the Middle Rio Grande looks like a black hole in the middle of a state where the water resource is more or less regulated. This fate is especially ironic when you consider that the Middle Rio Grande is both the longest-settled area of a deeply hydrologic state and the engine of future economic growth in the state.

Surface Water Management: Working within the Legal Framework

Kevin G. Flanigan, *Rio Grande Bureau, New Mexico Interstate Stream Commission*

There are six major reservoirs in New Mexico upstream of the Middle Rio Grande. This paper provides some background on how those reservoirs are operated within the current legal framework and how those operations meet various purposes and needs within the Middle Rio Grande.

Between the Colorado–New Mexico state line on the north and Elephant Butte Reservoir on the south, four major tributaries join the Rio Grande, including the Rio Chama, the Jemez River, the Rio Salado, and the Rio Puerco. The Rio Chama is the primary tributary, heading in the San Juan Mountains of southwest Colorado and joining the Rio Grande just north of Española. Other significant tributaries include the Red River, Rio Pueblo de Taos, Embudo Creek, and Galisteo Creek flowing out of the Sangre de Cristo Mountains; the Jemez River flowing out of the Jemez Mountains; and the Rio Salado and Rio Puerco, which join the Rio Grande just above San Acacia. With the exception of the Rio Chama and the larger streams originating in the Sangre de Cristos, these tributaries are ephemeral, flowing only during snowmelt runoff or in response to heavy precipitation events.

The six major reservoirs described here are Heron, El Vado, and Abiquiu on the Rio Chama; Cochiti on the Rio Grande; Galisteo on Galisteo Creek; and Jemez Canyon on the Jemez River. Reservoir storage is usually discussed in units of acre-feet, which is the amount of water that it takes to cover one acre to a depth of one foot, or approximately 326,000 gallons.

RIO GRANDE RESERVOIRS

Heron Reservoir

Heron Reservoir is located on Willow Creek just above its confluence with the Rio Chama in northern Rio Arriba County. It was constructed in 1971 with a storage capacity of 401,000 acre-feet and is owned and operated by the U.S. Bureau of Reclamation. Heron is the storage reservoir for the San Juan–Chama Project, a federally authorized diversion project that brings roughly 100,000 acre-feet per year of water across the continental divide from the San Juan River basin and into the Rio Grande basin. That water flows

El Vado Dam and Reservoir

through a series of tunnels into Willow Creek, then into Heron Reservoir. Heron is allowed to store only San Juan–Chama water; it is not authorized to store native Rio Grande water (water that originates as runoff within the Rio Grande basin). San Juan–Chama water is contracted to several different water users throughout the Upper and Middle Rio Grande including multiple municipalities, the Jicarilla Apache Nation, the Pueblo of Ohkay Owingeh (formerly San Juan), and two irrigation districts. The city of Albuquerque and the Middle Rio Grande Conservancy District are the two largest project contractors. San Juan–Chama Project water is managed and accounted separate from native Rio Grande water.

El Vado Reservoir

El Vado Reservoir is located on the Rio Chama just a few miles below Heron Reservoir. It was constructed in 1935 by the Middle Rio Grande Conservancy District (MRGCD) and has a storage capacity of 180,000 acre-feet. Both San Juan–Chama and native Rio Grande water are stored in El Vado. The U.S. Bureau of Reclamation currently operates El Vado primarily to provide supplemental irrigation supplies to the MRGCD by agreement with the district. Native water is stored pursuant to New Mexico Office of the State Engineer permit number 1690, issued in 1930 (at press time, ownership of El Vado Dam and Reservoir and state engineer permit number 1690 was the subject of a legal dispute between the MRGCD and the U.S. Bureau of Reclamation). The federal government also stores and releases water from El Vado Reservoir to the prior and paramount lands of the six middle Rio Grande pueblos during times of low flow on the Rio Grande. Those lands have senior water rights to any other MRGCD lands.

Abiquiu Reservoir

Abiquiu Reservoir is located on the Rio Chama approximately 30 miles downstream of El Vado, and about 30 miles upstream of the confluence of the Rio Chama with the Rio Grande. Abiquiu Reservoir was built in 1963, is owned and operated by the U.S. Army Corps of Engineers, and has a maximum capacity of 1,200,000 acre-feet at the top of the spillway crest. The reservoir was initially authorized as a flood and sediment control reservoir, but in 1981 Congress

Abiquiu Dam and Reservoir

authorized the reservoir to store up to 200,000 acre-feet of San Juan–Chama water. In 1988 Congress authorized Abiquiu to store up to 200,000 acre-feet of native Rio Grande water, provided that the storage space is not needed for San Juan–Chama water.

Cochiti Reservoir

Cochiti Reservoir was built in 1975, is owned and operated by the U.S. Army Corps of Engineers, and has a maximum capacity of 590,000 acre-feet at the top of the spillway crest. Cochiti is located approximately 50 miles upstream of Albuquerque and is the

Cochiti Dam and Reservoir

major flood control reservoir for the Middle Rio Grande valley. It is also the only reservoir on the mainstem of the Rio Grande above Elephant Butte Reservoir in New Mexico. Cochiti was initially authorized as a flood and sediment control reservoir. In 1964 Congress authorized the formation of a permanent recreation pool for Cochiti Reservoir of roughly 50,000 acre-feet, which is maintained with San Juan–Chama water.

Galisteo Reservoir

Galisteo Reservoir is located on Galisteo Creek approximately ten miles above its confluence with the Rio Grande near Santo Domingo Pueblo. Owned and operated by the U.S. Army Corps of Engineers, it was constructed in 1970 for flood and sediment control and has a maximum capacity of about 90,000 acre-feet at the top of the spillway crest. Galisteo is different from the other reservoirs in that its releases are uncontrolled below 5,000 cubic feet per second. There are no outlet control works, so what comes in essentially equals what goes out. Water becomes temporarily stored if inflow exceeds 5,000 cubic feet per second, so most of the time the reservoir is completely dry.

Jemez Canyon Reservoir

Jemez Canyon Reservoir is located on the Jemez River a few miles above its confluence with the Rio Grande. Owned and operated by the U.S. Corps of Engineers as a flood and sediment control reservoir, it was constructed in 1953 and has a maximum capacity

Galisteo Dam (at far left). This earth-filled structure is 2,820 feet long with a maximum height of 158 feet. There are no outlet control works, so water becomes temporarily stored only if inflow exceeds 5,000 cubic feet per second, and most of the time the reservoir is completely dry.

of about 100,000 acre-feet at the top of the spillway crest. Jemez Canyon is similar to Galisteo Reservoir in that it is also operated as a dry reservoir. However, flow out of the reservoir is controlled by outlet works that allow release of flood waters at a desirable rate.

All of the major reservoirs in the basin are operated in accordance with various federal and state laws that constrain or limit those operations to specific purposes or functions. The most important of these is the Rio Grande Compact.

THE RIO GRANDE COMPACT

The Rio Grande Compact, an interstate agreement that apportions the waters of the Rio Grande between the states of Colorado, New Mexico, and Texas was executed in 1938 and became effective in 1939. Under the compact New Mexico is allowed to consume on average roughly twice as much water as Colorado and three times as much as Texas. New Mexico's share includes the amount of water it is entitled to consume between the Colorado–New Mexico state line and the Otowi gage, the amount in the Middle Rio Grande valley between Otowi gage and Elephant Butte Reservoir (including all tributary inflow and San Juan–Chama Project water), and the amount in the Elephant Butte Irrigation District below Elephant Butte in the Lower Rio Grande.

There are a number of compact restrictions that have an impact on reservoir operations and surface water management in the Middle Rio Grande valley. The most important is Article VII, which prohibits increasing storage of native Rio Grande water in any upstream reservoir constructed after 1929 when the combined storage in Elephant Butte and Caballo Reservoirs, not including credit and San Juan-Chama Project water, is below 400,000 acre-feet. All of the major reservoirs are subject to this restriction except Heron, because it does not store native Rio Grande water.

The Article VII storage prohibition can have a major impact on water management in the middle valley, particularly on El Vado Reservoir, which primarily stores irrigation water for MRGCD. Article VII was invoked in 2002 for the first time since 1979 and was in effect until May of 2005. Since that time it has gone into and out of effect as water storage has fluctuated at Elephant Butte and Caballo Reservoirs. Article VII storage restrictions also impact McClure and Nichols Reservoirs on the Santa Fe River, two relatively small reservoirs with a combined capacity of slightly less than 4,000 acre-feet that provide a significant portion of the city of Santa Fe's water supply.

WATER OPERATIONS AND MANAGEMENT

The term "reservoir operations" refers to the rate and timing at which storage or inflow into a reservoir is released or detained. The term "water operations" includes downstream monitoring to ensure that desired flows are achieved from changes in reservoir operations, and management of downstream diversions of flows released from storage. There are essentially three main types of water operations that impact the Middle Rio Grande:

- Irrigation operations
- Flood control operations
- Environmental operations

The tools that are used by water managers to conduct these operations include near real-time flow and storage data provided by stream gages via satellite uplink, automatically controlled reservoir and diversion gates that can be supervised from the office, and sophisticated computer models to track water accounting and help plan operations.

One important thing to keep in mind while reading the following descriptions of specific reservoir operations is that very little of the native Rio Grande water originating within the basin is actually captured and stored in the major reservoirs. On average, roughly 100,000 acre-feet of native Rio Grande water, less than 10 percent of the annual average flow at Otowi gage, has been historically held in storage (at least temporarily) upstream of Elephant Butte. The vast majority of the combined storage of Heron, El Vado, Abiquiu, Cochiti, and Jemez Canyon Reservoirs has historically been San Juan–Chama Project water.

STORAGE AND FLOW

Reservoir storage and stream flow are intimately related. Flow can become storage by capturing it in some type of container, such as a reservoir. Storage can become flow by releasing it from that container. (A continuous flow of one cubic feet per second for 24 hours is equal to roughly two acre-feet of storage.)

Water is stored in reservoirs for several different purposes. Water stored for later release to meet a downstream demand, such as irrigation demand when stream flows naturally become low, is termed conservation storage because it is water conserved to meet a future use. The primary purpose of El Vado Reservoir is to provide conservation storage for irrigation use. Flood control storage is water temporarily stored to prevent or alleviate downstream flooding. Permanent storage, such as the Cochiti recreational pool, is maintained indefinitely to provide recreational, fish, and wildlife benefits.

WATER ACCOUNTING

All the water flowing through the basin is accounted in one fashion or another to ensure that its management and use is in compliance with all applicable law. All reservoir storage and flows at particular gages are accounted to ensure that Colorado is meeting its Rio Grande Compact obligation to New Mexico and that New Mexico is meeting its obligation to Texas. Water is also accounted on the level of individual ownership of various parties who have a right to its use such as the irrigation storage water released by the MRGCD, San Juan–Chama water moved from one reservoir to another by various parties, or supplemental water leased by the federal government for the endangered silvery minnow.

IRRIGATION OPERATIONS

Irrigation operations primarily consist of changing the rate and timing of storage releases from El Vado Reservoir to ensure there is sufficient flow in the Middle Rio Grande to meet the irrigation diversion needs of the MRGCD. To determine the rate of release, the MRGCD evaluates the amount of native flow moving downstream in the Rio Grande at Embudo and the amount of native flow contributed by the Rio Chama and other tributaries and compares that amount with their estimated future diversion demand. Diversion needs must be estimated two or three days into the future in order to determine how much storage to release from El Vado Reservoir to supplement the natural flow, as it takes that much time for those releases to reach the middle valley. Diversion needs fluctuate with weather conditions and the day of the week. Irrigation demand is generally higher on weekends, except holiday weekends. Irrigation storage is released only when the natural flow is insufficient to meet the MRGCD's irrigation needs. Natural flow is generally only sufficient to meet that need early and late in the irrigation season, during the snowmelt runoff and during periods of heavy monsoon activity.

FLOOD CONTROL OPERATIONS

Flood control operations adjust the rate and timing of releases or detention of inflow at the Corps of Engineers' flood control reservoirs: Abiquiu, Cochiti and Jemez Canyon Reservoirs. Releases at the fourth flood control reservoir—Galisteo—are uncontrolled. The four reservoirs are operated as a system to ensure that flow levels at critical downstream points are not exceeded. Flood control operations usually occur during snowmelt runoff when the mountain snowpack is heavier than normal and during heavy summer monsoon seasons. The snowmelt runoff of 2005 was the most recent major period of flood control operations, when approximately 75,000 acre-feet of flood control storage was detained in Abiquiu and 45,000 acre-feet in Cochiti. That water was released once runoff flows receded and it was safe to do so.

Article VII storage restrictions do not impact flood control operations at Abiquiu, Cochiti, or Jemez Canyon. In addition, in accordance with federal law, when the natural flow during the tail end of the snowmelt runoff drops to a level that is insufficient to meet MRGCD's diversion needs, any floodwater in storage is retained until after the irrigation season ends to ensure that the Rio Grande Project receives the water it would have if the flood control reservoirs did not exist.

ENVIRONMENTAL OPERATIONS

Environmental operations for the endangered silvery minnow have had the most impact on the Middle Rio Grande in recent years. Since 1996 the U.S. Bureau of Reclamation has been leasing water from willing parties to provide supplemental flows for the minnow in the middle valley. Since 2001 that supplemental water has been used to meet legally established levels of flow for the minnow as required by the Endangered Species Act. This water is leased and stored in Heron, El Vado, or Abiquiu Reservoirs and released during times when the natural flow of the river becomes too low to maintain certain levels in specific reaches of the Middle Rio Grande. A significant amount of management and coordination between the federal, state, and local water management agencies is necessary to successfully accomplish these operations. It is particularly difficult to efficiently provide relatively small flows to the lower end of the system at San Marcial by release of supplemental water stored in reservoirs on the Rio Chama when it takes five plus days for those releases to travel that distance.

CHAPTER THREE

CURRENT ISSUES ON THE MIDDLE RIO GRANDE

DECISION-MAKERS FIELD CONFERENCE 2007
San Acacia to Elephant Butte

Tributary to the Rio Grande near La Joya

Drought and Middle Rio Grande Water Management Issues

Ben Harding, *Hydrosphere Resource Consultants*

New Mexicans are a drought-tolerant species. This should come as no surprise. Earlier inhabitants of this land, the Ancestral Puebloans, survived using miniscule amounts of water, much of it carried by hand and stored in pots! Perhaps they even thrived; evidence indicates that the population of the Four Corners area was greater in the year 1000 than it was in the year 2000.

Modern New Mexico probably has no more water than it did one thousand years ago, but in those thousand years, technology and economy have allowed water to be stored in great volumes and moved great distances. Thus the modern resident of the Middle Rio Grande valley lives life with an abundance of water unimaginable to the Ancestral Puebloans. Even so, the state remains vulnerable to the adverse effects of drought. If New Mexico does not prepare for periods of drought, and then act properly, physical water shortages will be compounded by constraints on water storage arising from the Rio Grande Compact and the Endangered Species Act.

New Mexico has the cultural and historical perspective to embrace the reality of drought. I believe that the state also has the necessary will to take the difficult steps needed to prepare for drought. And New Mexico has many tools at its disposal with which to cope with drought.

THE NATURE OF DROUGHT

Probably the sparest meaningful definition of drought is a temporary period of water scarcity. Although the shortfall of snowfall or rainfall that brings about a water shortage is almost entirely outside our control, how we manage our water and natural systems over the long term can either mitigate or prolong and accentuate the effects of drought.

What are the adverse consequences of drought in the Rio Grande? The obvious consequence is a reduction in physical supply. Physical scarcity can imperil water supplies for irrigation or municipal uses, and water for natural systems and even aesthetics. As if that's not enough, a period of low stream flows can trigger parts of the Rio Grande Compact that place legal restrictions on the storage and release of water for use in the Middle Rio Grande valley, restrictions that can persist even after the drought has broken.

PHYSICAL SCARCITY

Physical scarcity is usually mitigated by storing water in plentiful years and applying it in water-short years. Both surface water and ground water storage are used to maintain physical supplies during times of drought. Increasing efficiency can help sustain supplies of stored water, but once stored supplies are depleted, continuing drought will inevitably result in a reduction in beneficial use and impacts on natural systems.

LEGAL SCARCITY—THE RIO GRANDE COMPACT

Legal scarcity can be painful. Even in times of abundant supplies, legal constraints such as those that might be imposed by the Rio Grande Compact on operation of reservoirs upstream of Elephant Butte Reservoir can restrict storage and later use of water. What is important to recognize about the compact is that even though it places strict constraints on how much Rio Grande water New Mexico is entitled to use over the long term, it provides substantial flexibility to the state in how it copes with drought over the short term. The compact sets out how much water New Mexico is entitled to consume in the middle valley in any given year. Article VI of the compact provides for a system of annual credits and debits, reflecting over- and under-delivery of water, respectively, in any one year, and it requires that a running sum of the annual debits and credits be maintained. So long as the running total accrues to an overall credit, there is no consequence, but once New Mexico accrues a debit then she must keep a reserve in New Mexico reservoirs in an amount equal to the accrued debit (Article VI restrictions). Article VI also sets an upper limit of 200,000 acre-feet on New Mexico's accrued debit. The Rio Grande Compact defines "Usable Water in Project Storage" as the amount of water that is legally available for release from Elephant Butte and Caballo Reservoirs. Article VII of the Rio Grande Compact restricts storage in some New Mexico reservoirs when

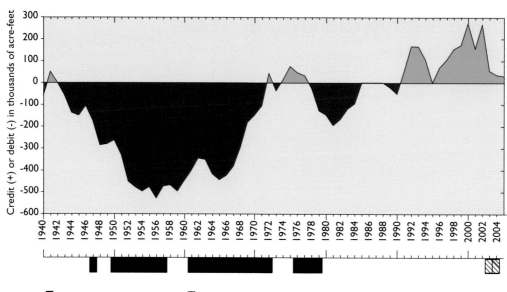

New Mexico's history of credits, debits, and Article VII restrictions. The red-colored areas in the upper chart represent debits resulting from New Mexico's under-deliveries under the Rio Grande Compact. Blue areas show credits resulting from accumulated over-deliveries.

The lower chart shows periods when Article VII restrictions were in place. The hatched bars show that in 2004 and 2005 restrictions would have been imposed but were offset by relinquishment of credits.

Usable Water in Project Storage falls below 400,000 acre-feet (Article VII restrictions). Under these circumstances, if New Mexico has accrued credits, it may relinquish some credits and store an equal amount of water. Enforcement of the Rio Grande Compact would be by a federal water master, by order of the U.S. Supreme Court.

CONSEQUENCES OF LEGAL SCARCITY

Compact restrictions on water storage reduce the effectiveness of upstream reservoirs and consequently reduce the physical supply of surface water available in the Middle Rio Grande valley. Once one fully understands the compact restrictions, one key fact becomes apparent: It is only when compact delivery credits are exhausted (either by overuse or by relinquishment) that any adverse legal consequences arise from the compact. A comparison of the situation during the current drought with the situation during the drought of the 1950s and 1960s illustrates this point.

THE CURRENT DROUGHT

Elephant Butte Reservoir spilled in 1995, and Usable Water in Project Storage stayed between 1.5 and 2 million acre-feet until 2000, at which point it began a steady decline. The cause of this decline is straightforward: Rio Grande Project users, in accordance with project rules, continued to get a full delivery from Elephant Butte Reservoir despite declining inflows. Usable storage slipped below the Article VII storage restriction trigger level in August of 2002, and stayed below that limit until the spring of 2005. Since that time it has bounced above and below the limit. At the end of 2002 New Mexico had an accrued credit of 265,000 acre-feet, so the state was able to offset the restrictions by relinquishing 175,000 acre-feet of credits, which allowed Middle Rio Grande water users to store a like amount of water in upstream reservoirs, primarily El Vado Reservoir but also in the city of Santa Fe's reservoir, as seen in the figure on this page. That 175,000 acre-feet of water, along with San Juan–Chama water, is the water that has been used over the past three years to supplement the river when natural supplies were scarce. It provided water to Middle Rio Grande Conservancy District farmers, to the citizens of Santa Fe, and for the Bureau of Reclamation to meet river-flow targets for the silvery minnow. Contrast this outcome with what happened in the 1950s.

THE DROUGHT OF THE 1950s AND 1960s

Compact accounting began in January 1940. The following five years were relatively wet; Elephant Butte

Reservoir stayed above one million acre-feet from 1941 through 1945 (it spilled in 1942), but, during this time New Mexico steadily accrued a debit, which, by the end of 1945, had reached 150,000 acre-feet. The debit continued to increase, exceeding the 200,000 acre-foot limit set in Article VI during 1948 and reaching a maximum of 529,000 acre-feet in 1956. New Mexico entered the severe drought of 1950–1956 with a deficit that already exceeded the limit set in Article VI. Not surprisingly, Texas sued New Mexico before the U.S. Supreme Court in the fall term of 1951, but this suit was thrown out on a technicality in February 1957. During the period from 1948 through 1968 New Mexico was continuously in violation of Article VI requirements and was under Article VII restrictions approximately two-thirds of the time. These restrictions amplified the impact of that already-severe drought.

Obviously nature deals us cards over which we have no control, but New Mexico can decide how to play its hand. Nothing about coping with drought will be easy, but careful preparation will help prevent or mitigate the most severe consequences of drought.

STRATEGY FOR THE LONG TERM

A strategy sets out big, long-term goals. I suggest three such goals for New Mexico:

- *Preserve stored water supplies.* There is not much more to be said about this strategy—it is the instinct of water managers. Remember that stored water includes ground water. Since San Juan–Chama Project water is accounted outside the Rio Grande Compact, preserving those supplies for times of shortage and compact restriction is vital.

- *Maintain an accrued compact credit.* Maintaining a compact credit avoids any legal restrictions on the use of reservoirs or stored water and thus is an adjunct to the first element of the drought-coping strategy. Because relinquishments may be necessary to offset Article VII restrictions, the accrued credit should be large enough to maintain a credit, even if small, after a relinquishment.

- *Develop contingency plans.* In the event that catastrophe strikes, have a plan in place, even if it is only a framework for decision making. Contingency plans should be developed for events that are precedented but extremely rare.

What constitutes "extremely rare" is a policy decision. Planning for ill-defined, unprecedented, or unforeseeable events should be done but falls outside the domain of drought. One example of an unprecedented event would be extensive wildfires that change the nature of the Rio Grande watershed in New Mexico and Colorado. The effects of climate change are nearly certain to occur, but exactly what those effects will be is ill-defined.

TACTICS FOR THE SHORT TERM

Tactics are the means of achieving strategic goals. In baseball, making outs is a strategy; throwing strikes is a tactic. I suggest three means to meet the strategic goals:

- *Increase efficiency of use.* Increasing the efficiency of agricultural and municipal use preserves physical supplies, either in reservoirs or aquifers. Although the amount of water consumed by a given agricultural or municipal use (beneficial consumptive use) is a stubborn fact, the amount of water applied to that use can be reduced by increasing efficiency.

- *Balance consumptive use against credit status.* The amount of New Mexico's accrued credit reflects the degree to which the state has balanced consumptive use against its entitlement under the Rio Grande Compact. If the state wishes to increase its credit (or reduce its debit), it must reduce consumptive use by reducing beneficial use, by reducing evaporation (primarily from Elephant Butte Reservoir), or by reducing consumptive use from riparian vegetation.

- *Balance credit status against project storage.* Article VII restrictions can occur even when New Mexico has properly balanced its water use against its entitlement (the current drought is an example). When project storage is high, it is sufficient to maintain only a positive credit. As Project Storage decreases, the probability of Article VII restrictions increases, and the accrued credits should be increased accordingly.

  The Rio Grande Compact also provides that accrued credits are reduced by the amount of spills. As project storage increases, the probability of a spill increases, and the state may wish to increase the amount of beneficial use, increase the amount of stored water (including recharge of ground water) to the extent possible, or both.

TOOLS

Administration. New Mexico is a prior appropriation state. This principle, which is set out in the New Mexico Constitution, means that a person who first puts water to beneficial use (and does so properly according to New Mexico law) will forever have the right to use water before those who put water to use later. A succinct statement of the prior appropriation principle is "First in time, first in line." Administration under the prior appropriation system is the big lever that the state possesses to balance beneficial consumptive use against New Mexico's entitlement. Although water has been put to use in the Middle Rio Grande valley for hundreds of years, water rights in the valley have not yet been adjudicated. Adjudication is a process wherein a court defines, once and for all, the priority and the quantity of each water right in the valley. Until the Middle Rio Grande is adjudicated, an interim approach, such as Active Water Resource Management, will be required for administration. It is worth considering that, as onerous as it may be to have the state engineer administering water rights in the Middle Rio Grande, administration by the state is preferable to administration by a federal water master.

Water bank. Strict priority administration of water rights is not economically efficient. Water markets provide one means of increasing economic efficiency but, because the process of transferring a water right is costly, existing water markets favor permanent transfers. In one sense a water bank is simply a set of rules that facilitates temporary, short-term transfers of water (as opposed to permanent transfers of water rights). A water bank, properly formulated, could reduce transaction costs and thereby facilitate temporary transfers of water. This would improve both market efficiency and the long-term prospects for irrigated agriculture in the Middle Rio Grande valley.

River management. Riparian consumptive use and reservoir evaporation (90 percent of which occurs at Elephant Butte Reservoir) combine to make up more than half of all depletions in the Middle Rio Grande valley. Reducing these depletions will not come easily—some programs to eliminate non-native vegetation have not lived up to their promise, but even modest reductions, accruing over the long term, will substantially improve the state's ability to maintain a compact credit.

Forecasting, statistics, and research. Good forecasts of water supply can improve water management decisions. Improved stream flow forecasts will become available, and targeted research can further improve forecasting in the Middle Rio Grande. Careful statistical analyses and modeling of historic and prehistoric (tree-ring reconstructions) stream flows can provide insight as to how to respond to a given forecast. Other fruitful areas for research are water markets, water banks, and approaches to reducing depletions from riparian vegetation and reservoir evaporation.

Information and education. Water management policies and practices must be transparent to all interested parties. As water management policies and practices evolve, it will be important to inform and educate all interested citizens to the facts and implications of those policies and practices.

A Killing "Cure"—Agricultural-to-Urban Water Transfers in the Middle Rio Grande Basin

Lisa Robert

In the summer of 2006 delegates from three adjacent water planning regions convened in Albuquerque to talk about the "gap" that exists between water supply and water demand in the central Rio Grande basin, and to address potential conflicts contained in their separate strategies for dealing with that shortfall. It was soon apparent that both the Jemez y Sangre and Middle Rio Grande planning regions had completed their state-mandated water plans in a hopeful vacuum: each assumed that agricultural water (upwards of 12,500 acres worth in the case of the Middle Rio Grande) could be acquired from somewhere downstream to help alleviate their respective insufficiencies.

For the most part, that "somewhere downstream" meant Socorro County, at the tail end of the Middle Rio Grande Conservancy District and just above Elephant Butte Reservoir, where, in terms of the interstate compact that apportions the river's annual flow, New Mexico ends and Texas begins. But planners in that bottommost region had a revelation for their neighbors to the north: "The idea of upstream regions coming to Socorro to transfer water is inconsistent with reality," said a Socorro/Sierra representative. "*We* have a deficit too. *There is no water to transfer.*"

Current conditions in the Middle Rio Grande basin are testing the old assumption that water can be transferred from agricultural to urban uses without jeopardizing either the relationship between river and ground water, or the one between ecology and economy.

Yet the export of historic water rights from rural Socorro to swelling urban centers upriver has been underway for many years. Within the Middle Rio Grande Conservancy District's Socorro Division, water appurtenant to more than 1,800 acres of farmland—nearly 25 percent of the senior rights once in existence between Lemitar and San Marcial—have already been sold, generally to facilitate development in Valencia, Bernalillo, Sandoval, or Santa Fe Counties. Agricultural water rights have also been transferred out of the valley north of Lemitar, and it is likely that well over 3,500 acres have been retired countywide. The costs of that loss may not be immediately apparent, but they are heavy, and they are certain.

Transfers are the mechanism by which water rights are moved from one category of use to another, and/or from one place of use to another within a stream system, in this case, the Middle Rio Grande basin. In the most common type of transfer, a surface water right is severed from the parcel of land it historically served and exchanged for the right to pump ground water in another part of the basin. This balancing act is necessary because (1) the surface water of the Rio Grande and ground water in adjacent aquifers is fundamentally one-and-the-same, and (2) the Rio Grande is a "fully appropriated" stream system, meaning there is no surplus to allocate to new uses. Retiring and transferring historic agricultural water rights offers the only ready source of supply for new municipal and industrial development. A transfer is therefore a reallocation of water, and by statute the state engineer is responsible for ensuring that a change in a right's point of diversion, or its place or purpose of use, will not impair another's water rights. In addition, the state is also mandated to consider possible impacts of the transfer on water conservation, and on "public welfare," although precisely what the latter means has yet to be determined. Notice of a proposed transfer must be posted in a newspaper of local, general circulation for three consecutive weeks, and the Office of the State Engineer may conduct a hearing on any subsequent protests.

But water transfers have not always been handled according to the foregoing rules. From the early 1970s to 1994, a policy known as dedication enabled an applicant to obtain a permit to appropriate ground water on the condition that senior water rights would eventually be retired to counter the effects of new pumping on the river. Agricultural lands from which the water would be transferred did not necessarily have to be identified; if they were identified, proof of the water right's validity was often nominal or entirely lacking; and no public notification process was required, meaning no protest of the transfer was possible. Worse, the requisite drying up of the transfer-from parcel was frequently deferred: Sellers were often allowed to continue irrigating until some future date, with no clear-cut method for ensuring that once the right was being fully exercised at the new location, water use would be terminated at the old location. In short, dedications evaded almost all of the legal requirements for transferring a water right.

The imprecise trail of middle basin water transfers is etched in pencil on a set of paper maps at the Office of the State Engineer. According to staff, the maps are accurate to within three weeks and thus comprise "the most complete" record available, yet they do not differentiate between dedications and formal transfers, and they do not indicate whether the transfer-from lands have actually been fallowed. Nor have all of these transactions been logged into WATERS, the state's GIS database, and data entry pertaining to the Middle Rio Grande has been temporarily halted. With no comprehensive inventory of water transfers, and no tally of "retired" parcels still being irrigated, it is virtually certain that present estimates of agricultural water rights in the mid-Rio Grande are erroneous. That miscalculation has the potential to affect every water right holder, every urban dweller, every resource agency, and every planning effort in the basin.

UNDERMINING THE FOUNDATIONS

Water law in New Mexico is based on priority. The state constitution and many historic treaties guarantee that the oldest appropriators of surface water—generally agricultural users—have seniority in time of shortage, and that newer uses will not be allowed to impair the exercise of older rights. Furthermore, the Rio Grande Compact unequivocally defines the amount of native water that can be consumed in the middle basin. But explosive population growth and development, environmental mandates, and the vagaries of climate change are stressing that finite supply. We are simply using more wet water than we receive, and we have been making up the difference with imported water (e.g., San Juan–Chama Project water) and ground water. Current calculations show a deficit of 40,000 acre-feet per year. In addition to that annual burden, the delayed consequences of past ground water pumping are just beginning to affect the surface flow of the river. The projected additional deficit due to ground water pumping is currently 71,000 acre-feet. Because the basin has not been adjudicated and the sum of its vested water rights is unknown, each new water use has the potential to intensify the regional deficit, impair senior water rights, and invite litigation. Uncertainty is inherent in every water transfer approved by the state engineer because such rulings may be reversed in a future adjudication. Likewise, a lawsuit prompted by compact debt, as occurred on the Pecos River in the 1980s, could cost New Mexico hundreds of millions and will without a doubt have to be paid in water, not dollars.

NEIGHBORHOOD TRAUMA

Water transfers reverberate throughout the community of origin, and the Socorro area, with its farm-based economy and public mandate to preserve agricultural tradition, is no exception. New Mexico acequias have long resisted water transfers on the basis of "third party effects," and those same arguments can be applied to any locale reliant on a fixed amount of irrigated land. As transfers occur, water conveyance system costs will be borne by fewer and fewer users, eventually threatening the practicality of delivery, and discouraging the cultivation of "marginal" lands. The loss of farm-associated revenue affects all local residents. According to Charles Howe, professor emeritus of economics at the University of Colorado-Boulder, and Christopher Goemans, a Ph.D. candidate in economics at the same institution, as agricultural acreage decreases, "activities linked to agriculture are negatively affected: Suppliers of agricultural inputs lose business, processors of agricultural outputs lose supply sources, financial institutions lose the demand for loans, etc." In human terms, those ripples reach far beyond the farmer who sells his water. Seed suppliers, equipment dealers, mechanics, field hands, 4-H kids, bankers, and bureaucrats will all feel the loss.

Agriculture forms the backbone of Socorro County's economy, but its total value is not reflected in crop census reports or income earned. Many Rio Grande valley farms are family-oriented, meaning those who raise the food also consume it. Given that reality, water transfers and the associated loss of agricultural land most certainly have the power to diminish local security and self-sufficiency. They also undermine the public welfare as defined by Socorro area residents, who have opted for rural living, not urban expansion, and who made the preservation of irrigated land a cornerstone of their regional water plan.

In addition to its agricultural riches, Socorro is home to several wildlife refuges and an increasing number of federal, state, and locally funded projects aimed at environmental restoration. Farmlands are integral to these programs because they provide a haven where diversity—tomorrow's saving grace—can flourish at little or no cost to the public. The attrition of agricultural land, as well as Socorro's position at the bottom of the water delivery system, jeopardizes restoration efforts already underway: The more water that is transferred out of the area and withdrawn upstream, the more difficult it becomes for the river and its proxy, the Middle Rio Grande Conservancy District, to transport sufficient flows to and through the Socorro reach.

Finally, agricultural water transfers facilitate "double and triple dipping," placing increased strain on an already-stretched resource. When a farm is sold with the intent of transferring its water to development outside the floodplain, the fallowed land is generally subdivided. Homes built on these properties are often served by domestic wells, ensuring that more water will now be used, some by new development outside the floodplain on either side of the river, and some by new houses in the valley. And of course those who purchase the former farmlands will want to water their acreage. To do so they might apply for an irrigation well, or they might drive one themselves without obtaining a permit, or they might pump more than they're entitled to from their non-metered domestic well, or they might reactivate an old ditch turnout and get water from the conservancy district. In every instance, the "fully appropriated" river is the source of supply, and is thus taxed with delivering several times the amount of the original surface right.

GROUND TRUTH AND PAPER

Throughout the West, and certainly along the populous Rio Grande, it has long been assumed that water for growth will come from agriculture. Implicit in that assumption is the belief that urban development is the highest and best use of Earth's most necessary resource. But is it?

Water transfers have consequences for the hydrologically dependent ecosystem that underpins agricultural and urban health alike. In recent years the creation of a regional water budget for planning purposes, substantial research into local river system dynamics on behalf of the bosque and the endangered silvery minnow, and an increasing ability to model the management of water in the basin have begun to reveal the crucial role played by irrigated land in the Rio Grande valley. Agricultural lands, in conjunction with the water delivery system of the Middle Rio Grande Conservancy District, function as a surrogate for the extensively altered river, helping to maintain the natural link between stream and ground water, giving surface flow access to its natural floodplain, and offering a form of aquifer recharge that is both practicable and economical. As this ecological role is undermined by the demise of irrigation, the entire system suffers a reduced capacity to deal with wet and dry extremes. What is being sacrificed is flexibility, perhaps *the* major key to surviving the uncertainties of global climate change.

Water reallocation will never produce the desired result of a balanced water budget. In shuffling promises, we hew to the path that created deficit in the first place. Transferring paper rights without understanding the wet-water price of such a philosophy endangers not just the so-called "place of origin," but the integrity and livability of the entire river basin.

Water in this desert place has the greatest value when it remains appurtenant to historic lands. There it retains standing under the law, embodies the very soul of New Mexico's traditional peoples, and performs a critical task in the health of the ecosystem. As diminishing oil reserves elevate the cost of transporting food and generating power, as the threat of terrorism counsels local independence, and as global corporations quietly buy up regional water supplies for commercial gain, we would be wise to safeguard the one irreplaceable asset that anchors us legally, defines us culturally, and sustains us environmentally.

I like to call dedications a "pump now, pay later" policy. Basically, you got to pump water out of storage and only acquire the water right when the flows in the river began to diminish. To me it's sort of like selling short—eventually you have to pay the bill. My big question is where are you going to get the rights to cover all this pumping? I tried to get a handle on the extent of these dedications and we came up with two or three different values. I can tell you that the number is so large it's probably going to require the majority of agriculture in the middle valley to change its purpose of use.
—Former State Engineer Tom Turney

A Tool for Floodplain Management along the Rio Grande

Matt Mitchell, *Rio Grande Agricultural Land Trust*
Dick Kreiner, *U. S. Army Corps of Engineers (retired)*

There are a number of issues in the reach of the Rio Grande from San Acacia to San Marcial that threaten future river health and its human and non-human inhabitants. Non-native plants like salt cedar and Russian olive have choked the river channel, restricting high flows of water and limiting regeneration of native plants. These invasive species also pose an extreme fire threat to residents of the area. There is currently no zoning in the floodplain in this reach, and development in flood-prone areas is a potential threat to river management. Incising of the river has decreased the number of regenerative overbank flood events, and, as a result, the aging cottonwood/willow forest is not being replaced. Finally, the San Marcial railroad bridge, because of silt deposition, now has limited capacity to pass seasonal flows of water. Like development, this issue is a limiting factor for water managers upstream. If these issues are not addressed, the river system will continue to degrade.

This paper is a look back over the last fifteen years at activities dealing with floodplain management (or the lack thereof) in Socorro County below San Acacia Diversion Dam. During most of those years, Dick Kreiner was chief of the Reservoir Control Section of the Albuquerque District, U.S. Army Corps of Engineers, and responsible for overseeing flood-control operations at corps reservoirs. It was the corps' job to determine how much water could be released safely from upstream corps dams, and to develop and implement the appropriate operating criteria for day-to-day operations. If an individual had a problem with our operations, they would contact the Corps of Engineers and their situation would be evaluated.

This was the case in 1992, when the corps was working with the Bureau of Reclamation and other agencies to see if they could put a little water in the overbank areas along the Rio Grande for the benefit of the riparian community we call "the bosque." Around this time everyone was beginning to realize that the bosque within the levees was in need of periodic overbank flows to maintain the health of the riparian vegetation. Cochiti Dam, 50 miles north of Albuquerque, had been in operation since 1975 and had quite effectively controlled the high flood flows coming into the Middle Rio Grande valley. It also had cut off most of the sediment load the river was carrying. With the sediment cut off, the river began to scour a deeper channel, and overbank areas in the northern section of the Middle Rio Grande were becoming isolated. It was taking higher flows to wet overbank areas that used to get flooded before construction of Cochiti Dam. The southern sections of the Middle Rio Grande, from Isleta Pueblo to Elephant Butte Reservoir, have areas that flood and therefore have a healthier bosque.

In 1992, when word was sent out about higher flows coming down the Rio Grande to help the environment, an individual contacted the Bureau of Reclamation and said, "Hey, I'm down here in Bosquecito, and you're going to flood my home if you increase the flow in the Rio Grande." Sure enough, after a quick trip down to Bosquecito, there was this new house right along the bank of the river. There are more than 13,000 square miles of uncontrolled area below Cochiti Dam, most of it in the Rio Puerco and Rio Salado drainages. Who, in their right mind, would build a house next to the river with the potential for flooding being so high? As it turned out, the house was built during a dry year and was later sold to this individual. To make a long story short: A small dike was constructed around the house, and the higher test flows followed with no damage to the home.

These higher test flows were timed to mimic the historic high flows on the Rio Grande and designed to promote native cottonwood and willow establishment. A project on the Bosque del Apache National Wildlife Refuge in 1993 and 1994 showed that if higher flows were timed right, native trees could become established and out-compete salt cedar for space on the floodplain. Fourteen years after the test, the native trees established back then have formed a cottonwood forest that still keeps salt cedar out. These test flows and habitat restoration projects started a long series of activities to figure out how to keep this floodplain open for flood waters, how to promote more native plants (instead of salt cedar), and how to assist landowners with improvements on their land.

Most of the Rio Grande farther north in Valencia and Bernalillo Counties is confined by levees on both sides, and the levees pretty much keep out development. In Socorro County below San Acacia Diversion Dam the levee is only on the west side of the river. The east side of the river has large flood-prone areas of private property north of Bosque del Apache National Wildlife Refuge. Shortly after the higher flows of 1992 Army Corps of Engineers representatives briefed the Socorro County Commission about this situation and offered assistance to them if they wanted to pursue participation in the National Flood Insurance Program. The county commission was also advised that for the safety of their residents they might want to consider zoning the river corridor to prevent further development in the Rio Grande floodplain. They thanked the corps for their time and went on to the next item on the agenda. It was very obvious to the corps that there was no political will to zone private property along the river.

What followed was a very interesting series of discussions on how to preserve these flood-prone areas. Advisors from Bosque del Apache National Wildlife Refuge, the Corps of Engineers, and other agencies helped concerned local citizens come up with a potential solution, one that would keep the flood-prone portions of private lands on the east side free from development; would remove the salt cedar and replace it with native grasses, shrubs, and trees; and (most important) would keep it in private ownership. The first meeting of this informal group, which came to be known as the Floodplain Management Group, occurred on the first day of spring in 1999. At that meeting in Bosquecito, the agencies present and the Save our Bosque Task Force got the thumbs up to pursue a floodplain management and habitat restoration program based on the ideas listed above.

The reason there had not been a big problem with development along the east side of the river wasn't because people were afraid of being flooded; they were afraid of getting burned out. Most of this reach was infested with salt cedar, and fires periodically raced through the bosque and adjoining lands with a terrorizing affect on the residents. What if we found a way to remove and control the salt cedar for the landowner in exchange for a conservation easement that prohibits development on the portion of their land in the floodplain? To sweeten the deal, restoration of native plant species could be included to provide wildlife habitat and increase the monetary value of their open lands.

The Save Our Bosque Task Force, the Bosque Improvement Group, several land trust organizations, and the local private landowners worked together to get a working program started. The Rio Grande Agricultural Land Trust received funding from the Bosque Improvement Group to do outreach work with the goal of educating and informing landowners about conservation easements. The Save Our Bosque Task Force received funding from a number of federal, local, and non-profit organizations to complete a feasibility study and conceptual habitat restoration plan for the valley. Both of these documents were done by 2004. One strategy is that the value derived from the retiring of development rights through a conservation easement can be used by the landowners for the required match to obtain state and federal dollars for habitat restoration work in the floodplain. The land remains in private hands, and the landowner gets a long-term partner in the habitat restoration on their lands. After contacting most of the landowners in the area, eight families expressed an interest right away in enrolling in such a program. Many others took a positive but "wait and see" position. Fundraising by the Rio Grande Agricultural Land Trust and the Save Our Bosque Task Force continues, and some dollars have been received to pay for these preliminary projects. A North American Wetlands Conservation Act grant was obtained to fund the establishment of six easements, and habitat restoration dollars have been received through the U.S. Fish and Wildlife Service, with other potential sources contacted. Two of the landowners will be ready to finalize their conservation easements in early 2007, with the others to follow when funding for habitat restoration is made available.

It is important to point out that the priority of those working on this effort was to solve floodplain development problems at the local level. The adage "think globally, act locally" comes to mind. One ill-advised home built on the bank of the Rio Grande has the potential to alter flood-control operations at federal dams that protect nearly a half million people. It also has the potential to jeopardize operations that are striving to sustain and enhance thousands of acres of riparian wetlands, forests, and grasslands, and their associated wildlife communities. These periodic flood waters keep the river channel open and more able to handle flood water that might otherwise threaten the levee protecting farms and communities on the west side of the river.

In the context of the short history provided above, it can be said that there is really no comprehensive program for addressing floodplain management in the reach. Because of the extensive private land in and around the floodplain, workable solutions must begin at the local level to be successful. Unregulated runoff from the Rio Puerco and Rio Salado into the Rio Grande above San Acacia continues to be a major flooding threat to downstream residents.

The Save Our Bosque Task Force, the Rio Grande Agricultural Land Trust, and concerned citizens will continue to work toward solving these important problems for area residents and all New Mexicans, with the voluntary help of the landowners on the east side of the Rio Grande. These landowners have shown their respect and love for their lands and for the Rio Grande, and they will be the best stewards for the future.

Salt Cedar Control: Exotic Species in the San Acacia Reach

James Cleverly and Gina Dello Russo, *U.S. Fish and Wildlife Service*

Although there are a number of exotic or invasive plants in the Middle Rio Grande, salt cedar is considered a risk to overall ecosystem health because it forms dense monotypic stands, poses a fire danger to adjacent plants, and has limited benefit to wildlife species. Salt cedar has infested widespread areas of the southwestern United States. In locations where salt cedar has proliferated, accounts of dried-up wetlands and saline soils are told. These plants are native primarily in Asia and the Middle East, so salt cedar is well adapted to our deserts' hot and dry climate. Even so, salt cedar responds to excess moisture with enhanced growth and water use. Due to alleged extravagant water use and observed extreme fire hazard of salt cedar thickets, restoration of the Rio Grande bosque through removal of salt cedar and other invasive vegetation is being relentlessly pursued.

Salt cedar, including several species of the genus *Tamarix*, is one of the most prolific plant species found in the San Acacia reach of the Middle Rio Grande. Infestations can be especially dense in this reach, where impenetrable monospecific stands are common. Downstream from the outflow of the Rios Puerco and Salado, summer flooding provides ideal conditions for germination of salt cedar's many seeds at a time when none of the native cottonwood and willow trees produces seeds. Once established, salt cedar has a further advantage in the San Acacia reach, where the areas of deep water table make it difficult to maintain vigorous native forests. Extraordinary variations in ground water levels are normal in the San Acacia reach, permitting salt cedar thickets to consume excessive amounts of water under some conditions. At the high density that some salt cedar stands reach—more than 6,000 plants per acre—water loss has been estimated in the range of 200 gallons per plant per year—or about three and a half acre-feet of water per acre of vegetation. Research has demonstrated that the water lost from salt cedar leaves can be one and a half acre-feet greater than water lost from an acre of water-conserving native vegetation. However, salt cedar infestations are notoriously difficult to remove effectively, and new growth from buried root crowns can actually consume more water than the salt cedars that were removed in the first place.

IN MONOTYPIC STANDS

Many salt cedar control techniques have been developed and applied in the Middle Rio Grande. Over the past twenty years, mechanical control on the Bosque del Apache National Wildlife Refuge and other locations has been done using bulldozers equipped with root plows or root rakes. This technique is used in areas of dense monotypic stands where other native

Schematic of above-ground and below-ground structure of *Tamarix* species, commonly known as salt cedar.

vegetation will not be impacted. On the wildlife refuge, effective control is achieved when salt cedars are reduced by 99 percent. The last raking should occur in the hottest part of the summer to assure drying of the remaining exposed roots. This technique is not applicable in areas with a high ground water table or in other areas where heavy machinery can't operate.

Root plow attached to a D7 dozer, sheering root crowns of salt cedar at the Bosque del Apache NWR, after the above-ground part of plant has been removed.

In those areas, aerial herbicide application has been used for initial control, followed by a controlled burn or mulching of the standing dead tree limbs. Follow-up spraying, grazing, or mulching is always necessary to get close to 99 percent control.

IN MIXED STANDS

In areas with both salt cedar and native plants a gentler approach is necessary. Efforts underway to remove salt cedar from cottonwood forests focus on reducing the fire danger to the native trees and surrounding communities. The Save Our Bosque Task Force is a 501(c)3 non-profit organization formed to 1993 to work toward sustainable riparian areas in the San Acacia reach. The task force has taken the lead in establishing fuel breaks near communities and native forests. The work continues with larger projects planned for the east side of the river near the communities of Bosquecito, Pueblito, and San Pedro. Masticators, excavators, chain saws, mowers, herbicide, and goats have all been used in these mixed stands. And experimentation continues on salt cedar control techniques for unique situations. Whether in solid or mixed stands of salt cedar, there is a necessary follow-up treatment on the remaining live shoots. Current techniques include excavator root extraction, herbicide spot spraying, mowing, or grazing. These follow-up techniques have been tried in various places with differing success. Whether in dense or mixed stands of salt cedar, initial control is only the first step to long-term maintenance of these areas.

Other Prevalent Invasive Species in the San Acacia Reach

Field Bindweed
A creeping perennial that forms dense mats, has a deep root system, and produces seeds with a long viability (up to 50 years). It occurs in most every yard and field in the floodplain in Socorro County. Control can be through herbicide treatment or continual disking.

Camelthorn
A spiny, creeping perennial with greenish stems and slender yellow tipped spines. It has a spreading root system, and is found on ditchbanks, roads, and pastures. It is just coming into Socorro County and has been found in the historic floodplain in San Antonio. Control can be through hand digging and herbicide treatment.

Russian Knapweed
A creeping perennial that forms large colonies, spreading from root buds. It is 3 to 4 feet tall, has deep roots (up to 25 feet), and lavender flowers. It occurs in both the historic and active floodplain of the Rio Grande in Socorro County. Control is usually through herbicide treatment.

Perennial Pepperweed
A creeping perennial that forms large colonies and can reach 6 feet tall. The largest concentration in New Mexico occurs on the Rio Grande. It spreads easily and now occurs on ditch banks, in fields, and along the river. It can grow under native and invasive trees and is hard to control, requiring multiple treatments with herbicide.

Russian Olive
An ornamental tree that grows to 30 feet; has red berries in fall, and large spines. This tree was brought into the area as a fast-growing windbreak and bank stabilizer. It has become established on the river mainly through seed dispersal by birds. It forms dense stands along the river, armoring the banks and limiting other native plant growth.

Other invasive or non-native trees found on the Middle Rio Grande include Siberian elm, tree of heaven, and mulberry. All of these invasive plants compete with native plants for room to grow. They can be effectively controlled if follow-up rehabilitation includes establishment of competitive native plants and continual monitoring for re-infestation. Important long-term controls for invasive plants should include addressing seed sources from adjacent areas or upstream on the main stem and on tributaries that feed into the river. Care must be taken when using seed or mulch mixes so that invasive seeds are not included in the mix.

This 1999 aerial photo shows the dense salt cedar at the south end of the Bosque del Apache NWR. The south boundary is at the top of the picture. Dark vegetation is monotypic stands of salt cedar, lighter vegetation is cottonwood forest. Panels visible in the lower left of the photo are part of a salt cedar control experiment area, testing mechanical removal and chemical treatments of salt cedar.

FOLLOWING SALT CEDAR CONTROL

Planting or promoting the natural regeneration of native vegetation, including trees, shrubs, and grasses, is critical for long-term management of these floodplain lands. The native plants can compete quite effectively with salt cedar and other undesirable plants once established. Techniques for this part of effective salt cedar management are being developed. Research into appropriate grass species for different soil textures and ground water levels is in progress. Information on native tree and shrub planting requirements and analysis of the resulting quality of planted areas is also being gathered. As an example, when salt cedar, sometimes used by the endangered southwestern willow flycatcher in this and other river systems, is removed, willow plantings can provide new and improved habitat. Experiments to improve cost effectiveness in establishing willows and enhancing flycatcher habitat are underway on three private and public lands habitat restoration projects this year.

There is an expanding body of knowledge of how to effectively control salt cedar and a growing interest in doing so. But achieving the goal of removing salt cedar and other invasive species such as Russian olive for the purpose of reduced fire hazard, improved biological diversity, or water salvage remains a challenge. Monitoring a site both before and following restoration provides independent evaluation of the pitfalls and successes of a restoration project. For example, the University of New Mexico has conducted continuous monitoring of an understory restoration project in Albuquerque's South Valley, where salt cedar and Russian olive shoots were mechanically removed, and the remaining stumps were treated with herbicide. At this site plant water use was reduced by 21 percent, or 0.85 acre-feet per acre restored, during the first year following restoration. Even though the Russian olive at the site re-sprouted immediately following herbicide application, growth was insufficient to affect the water budget. However, salt cedar re-sprouted from the remaining stumps at the outset of the second year,

This is a 2003 photo of the same area after aerial spraying; lines have been cut for a topographic survey of the area. The view is to the east toward areas that have had the dead (chemically treated) salt cedar burned off.

negating further water savings and illustrating the importance of return visits in removing salt cedar in a manner that succeeds in restoration goals. Without monitoring this site, we are destined to continue spending money on insufficient and ineffective restoration attempts.

This photo of the same area was taken in December 2006 after the Marcial Fire (June 2006) and following root plowing and root raking. The focus area is in the center left of the photo. Note the cleared areas to the south and the changes visible since the 1999 photo.

LESSONS LEARNED

Some things that salt cedar control and/or habitat restoration practitioners have learned are that each site is unique in terms of the suite of control and restoration techniques that will work. This is because sites have unique ground water/surface water connection, ground water levels, soil chemistry, soil texture, and flooding potential. Each site is also unique in terms of the existing water use by vegetation and the potential water savings or cost with restoration. Many research projects have measured site specific water use in both native and invasive forests, and the accuracy of those measurements is improving. We generalize the water use by different plant communities at this point, make assumptions, and at some point will have to decide, when we have reached the accuracy we need, just what we can expect to achieve in such a diverse system. The Save Our Bosque Task Force looked at the potential for water savings in the San Acacia reach from the perspective of change in plant density, assuming there would be a change in "leaf area index," one measure that has been correlated with water use in recent research experiments. Using the change in plant density through habitat restoration, and with the assumption that as you move away from the river channel and gain elevation, you are slowly disconnecting from the river and the shallow ground water of the river basin, a water savings by the natural system was achieved. Improved biological diversity and decreased fire danger from dense vegetation were also achieved. What is most important to remember is that these improvements to the system happen over a long period of time. Diverse native riparian wildlife habitat is only sustainable if the river is allowed to exist as a dynamic part of the system.

Private landowners from the San Acacia reach villages are very interested in salt cedar control and habitat restoration projects that result in fire protection for their homes and improved wildlife habitat on their lands. They also are aware of the need to control salt cedar over the long term so their lands don't become salt cedar thickets once again.

There are four requirements for success in long-term effective salt cedar control and habitat restoration:

- Select replacement native vegetation appropriate to the site's flooding potential, ground water level, and soil conditions

- Understand that healthy river functions—sediment movement, occasional flooding to increase decomposition of plants, maintaining an open channel, and ground water recharge—are necessary for cottonwoods and willows to survive and thrive next to the river

- Support landowners for occasional follow-up salt cedar control treatments

- Prioritize, coordinate, and collaborate with other agencies and entities on large-scale, long-term projects to assure funding and information sharing

In this way we work together toward the goals of reduced fire danger, improved biological diversity, and efficient water use by an improved natural system.

To meet these four requirements for success, good working relationships have been developed among private landowners, land managers, federal and state government agencies, non-government agencies, and the Save Our Bosque Task Force in the San Acacia reach. Over the past 13 years, the task force has completed a feasibility study and a conceptual habitat restoration plan for the valley. This plan provides important information on flooding potential, existing vegetation, and possible restoration techniques. From this planning effort the Save Our Bosque Task Force and other interested stakeholders have developed the "Floodplain Management Program" for land protection and habitat restoration. This program seeks long-term solutions to issues of floodplain encroachment, protecting and enhancing private property values, and improving water management flexibility through the San Acacia reach. The Floodplain Management Program is underway and has attracted the interest of a number of landowners.

The Endangered Species Act and the San Acacia Reach

Jennifer M. Parody, *U.S. Fish and Wildlife Service*

New Mexico's longest river, the Rio Grande, supports three of the state's "listed" species: the endangered Rio Grande silvery minnow, the Southwestern willow flycatcher, and the threatened bald eagle. Each of these species has unique habitat requirements found within the Rio Grande ecosystem. The silvery minnow and flycatcher, in particular, have been affected by past large-scale water operations and management. Both have benefited from recent collaborative efforts to protect existing populations, restore habitat, and manage river flows throughout the system. The San Acacia reach holds special challenges and opportunities for these species and over the long term may be the key to their recovery.

RIO GRANDE SILVERY MINNOW

The Rio Grande silvery minnow historically occupied close to 2,400 river miles in New Mexico and Texas. It was found in the Rio Grande from Española down through Texas to the Gulf of Mexico. It also occupied the Pecos River, from Santa Rosa downstream to its confluence with the Rio Grande in Texas.

The Rio Grande silvery minnow.

Currently the silvery minnow is found only in what is known as the Middle Rio Grande of New Mexico, a 174-mile stretch of river that runs from Cochiti Dam to the headwaters of Elephant Butte Reservoir, about 7 percent of its former range. The silvery minnow was listed as endangered in 1994.

High-quality habitat for the silvery minnow includes stream margins, side channels, and off-channel pools where water velocities are low. Stream reaches dominated by straight, narrow, incised channels with rapid flows typically are not occupied by silvery minnow. The species is a pelagic (open water) spawner that produces 3,000 to 6,000 eggs during a single spawning event. Adults spawn in late spring and early summer (May to June) in association with spring runoff. Eggs and larvae remain in the drift for three to five days.

Approximately three days after hatching, the larvae move to low-velocity habitats where food (mainly phytoplankton and zooplankton—microscopic plants and animals) is abundant and predators are scarce. Higher flows that move water out of the channel and into the floodplain help transport eggs and larvae to nursery habitat. In the winter, silvery minnows congregate in deep, slower waters near debris piles and submerged vegetation.

SOUTHWESTERN WILLOW FLYCATCHER

The southwestern willow flycatcher is a small songbird that winters in the neotropics (southern Mexico to South America) and breeds in the southwestern United States. It was listed as endangered in 1995, largely due to habitat loss and degradation. The highest concentrations of flycatchers in New Mexico are on the Gila River near the town of Cliff, and on the Rio Grande in Socorro County at the headwaters of Elephant Butte Reservoir.

The flycatcher nests in dense riparian areas along rivers, streams, or other wetlands. Nest sites are dominated by dense growths of willows, seepwillow, or other shrubs and medium-sized trees. There may be an overstory of cottonwood, tamarisk, or other large trees, but this is not always the case. In some areas, the flycatcher will nest in habitats dominated by tamarisk and Russian olive. One of the most important characteristics of the habitat appears to be the presence of dense vegetation, usually throughout all vegetation layers. Almost all flycatcher breeding habitats are within proximity (less than 20 yards) of water or very saturated soil. This water may be in the form of large rivers, smaller streams, springs, or marshes. At some sites, surface water is present early in the nesting

Southwestern willow flycatcher.

season, but the ground gradually dries up as the season progresses. Ultimately, the breeding site must have a water table high enough to support riparian vegetation.

THE RIO GRANDE PAST AND PRESENT

Before widespread human influence the Rio Grande was a wide, shallow, perennially flowing river with a shifting sand bottom. The river freely migrated across a wide floodplain. This floodplain was composed of many secondary channels, backwaters, lakes, and marshes. Floods maintained a high water table that provided some open water during very dry times. Such an environment was ideal for supporting silvery minnows and flycatchers.

The Rio Grande, however, has undergone considerable change in the last 150 years, and it is no longer the highly dynamic system it once was. Dams and irrigation diversions are operated primarily to reduce flooding and to supply water for irrigation. In many areas, channel incision has reduced overbank flow onto the floodplain. In the San Acacia reach drying is common. These factors represent threats to the long-term survival of the silvery minnow and flycatcher on the Rio Grande. But recent efforts to restore the Rio Grande and protect these endangered species are reducing threats and improving the silvery minnow and flycatcher's chances of recovery.

THE ENDANGERED SPECIES ACT COLLABORATIVE PROGRAM AND THE 2003 BIOLOGICAL OPINION

The Middle Rio Grande is being protected and restored through the efforts of many organizations and entities. Of particular prominence, both due to its responsibility and membership, is the Middle Rio Grande Endangered Species Act Collaborative Program. Created in 2000 as the "ESA Workgroup" in response to litigation and conflict over water/endangered species issues, the Collaborative Program now includes more than twenty active signatories including state and federal agencies, local and tribal governments, universities, and farming organizations. The program has three interrelated goals:

- To meet the requirements of the Endangered Species Act
- Provide water to those who hold valid water rights
- Comply with the obligations of the multi-state Rio Grande Compact

One of the main responsibilities of the program is to help implement the U.S. Fish & Wildlife Service's programmatic 2003 Biological Opinion on water operations issued to the Bureau of Reclamation and the Corps of Engineers under Section 7 of the ESA. This opinion evaluates the effects of all contractual water deliveries and other operations of the river including river maintenance and flood control (the proposed action), identifies strategies to alleviate jeopardy to listed species, and provides "incidental take" coverage. All federal and non-federal parties that divert water from Cochiti Dam to Elephant Butte Reservoir are afforded ESA coverage, including incidental take, under the 2003 Biological Opinion. This overarching legal protection provides a strong incentive for all partners to assist in meeting the requirements of the 2003 Biological Opinion, and for participation in the Collaborative Program.

The 2003 Biological Opinion determined that proposed diversions and river management actions were likely to cause jeopardy to the silvery minnow and flycatcher, and provided a reasonable and prudent alternative with multiple elements. The alternative requires:

- Coordinated water operations and minimum river flows
- Habitat restoration
- Population management
- Improvements to water quality

The Collaborative Program receives, on average, $10

million annually, through a congressional earmark, to implement these activities. Cost sharing provided by the State of New Mexico has been critical to demonstrating to Congress the program's multi-agency representation and to ensure a non-federal voice in decision making. Through cooperative management, the Collaborative Program has been successful in meeting flow requirements for the past four years and in restoring several hundred acres of habitat. By augmenting flows and with two years of record-high snowmelt (2004 and 2005), silvery minnow populations are returning to prelisting levels. Flycatcher populations have also been stabilized, and its habitat is being created throughout the Rio Grande.

Constructed embayment in the Albuquerque reach of the Rio Grande, designed to provide slow water, nursery habitat for the Rio Grande silvery minnow.

THE SAN ACACIA REACH

The San Acacia reach presents unique challenges and opportunities for silvery minnow and flycatcher protection. This reach encompasses 56 of the 174 river miles within the Middle Rio Grande. This represents 32 percent of the occupied range of the silvery minnow and some of the highest quality habitat found in the system. The largest concentration of flycatchers is also found in this reach. Of the 174 flycatcher territo-

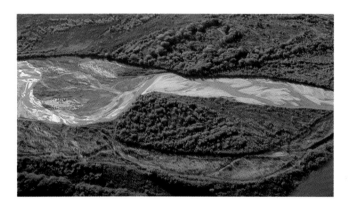

Aerial view of the Rio Grande in the Bosque del Apache National Wildlife Refuge, showing mature cottonwood bosque, wetland habitats, and some channel braiding, all indicative of dynamic river processes within a connected floodplain.

ries found on the Rio Grande in 2005, 110 were located south of San Acacia diversion dam. Currently the best and largest contiguous habitat area for flycatchers along the Rio Grande is south of San Acacia near San Marcial.

Generally, habitat on the Rio Grande for both species tends to increase in quality from north to south. Although poor quality (high velocity, channelized) areas and good quality (sand bars, back channels, slackwater) habitat can be found in all reaches, the San Acacia reach exhibits the greatest degree of river/floodplain connectivity and channel complexity. It also contains the largest number of riverine wetlands. What makes this reach challenging, however, is its tendency to dry due predominantly to diversions, ground water pumping, and river drainage to the Low Flow Conveyance Channel.

The Low Flow Conveyance Channel follows the river for 75 miles. It was designed in part to expedite delivery of water to Elephant Butte Reservoir during low flow conditions, as required by the Rio Grande Compact. Water was diverted to the Low Flow Conveyance Channel from the Rio Grande from 1959 to 1985. Because the Low Flow Conveyance Channel is at a lower elevation than the river bed, there is seepage from the river to the Low Flow Conveyance Channel. This causes a significant loss of surface flows in the river channel.

The San Acacia reach has experienced significant drying almost every summer since the mid-1990s. This strains resident and migrant populations of fish and wildlife and challenges efforts to maintain and restore habitat. Low flows and a lack of consistent overbank flooding can negatively affect riparian vegetation and increase fire danger. Not only does this vegetation rely on a high water table, but flood flows remove flammable understory debris, maintain a safe channel capacity, and create space for young plants to grow.

THE RIO GRANDE OF TOMORROW

To succeed in protecting the endangered species of the Rio Grande, we must begin by returning to the Rio Grande its ability to renew itself and its habitats.

Multiple restoration techniques are available to increase river dynamics and complexity. Most of these encourage lateral river movement within the confines of flood control levees, increase river/floodplain connectivity, and create habitats that may be inundated at lower flood flows. Such techniques include lowering banklines, modifying in-channel islands, reconnecting isolated channels, and building embayments. Broad application of these techniques can increase the amount of habitat available for both the silvery minnow and flycatcher. If strategically located, suitable habitat may be supported even during times of lower flows. Significant potential exists for such habitat improvements in the San Acacia reach. Due to a lack of levees on the east side and an already high degree of floodplain connectivity, many areas within this reach may be easily reconnected to the riverbed.

OPPORTUNITIES TO PARTICIPATE

Multiple opportunities exist for those interested in participating in endangered species habitat restoration, monitoring, and research. The Collaborative Program provides funds each year for projects and activities throughout its program area on the Rio Grande (http://www.fws.gov/mrgesacp/). Funding for projects that benefit listed and non-listed species is available through:

- The Middle Rio Grande Bosque Initiative (http://www.fws.gov/southwest/mrgbi/)

- The Partners for Fish and Wildlife Program (http://www.fws.gov/ifw2es/newmexico/)

- Tribal Wildlife and Tribal Landowner Incentive Programs and other grant opportunities may be found at http://www.fws.gov/grants/

Suggested Reading

Middle Rio Grande Ecosystem Bosque Biological Management Plan; the first decade—a review and update, Robert, L., Aurora Publishing, Albuquerque, New Mexico, 2005.

Habitat restoration plan for the Middle Rio Grande, Tetra Tech EM Inc., 2004.

Biological and conference opinions on the effects of actions associated with the programmatic biological assessment of Bureau of Reclamation's Water and River Maintenance Operations, Army Corps of Engineers' flood control operation, and related non-federal actions on the Middle Rio Grande, New Mexico, March 17, plus amendments, U.S. Fish and Wildlife Service, 2003.

A Few Definitions

Listed Species—Under the Endangered Species Act (ESA), species may be listed as threatened or endangered. Endangered means a species is in danger of extinction throughout all or a significant part of its range—the geographic area a species is known to or believed to occupy. Threatened means a species is likely to become endangered within the foreseeable future throughout all or a significant part of its range. The purpose of the ESA is to protect and recover these imperiled species and the ecosystems upon which they depend. The U.S. Fish and Wildlife Service is the agency principally responsible for administering the ESA.

Section 7 Consultation—Section 7 of the ESA requires federal agencies to use their legal authorities to promote the conservation purposes of the law. This section also requires federal agencies to consult with the U.S. Fish and Wildlife Service to ensure that actions they authorize, fund, or carry out will not jeopardize listed species. The consulting agency then receives a "biological opinion" on the proposed action. In the relatively few cases where the U.S. Fish and Wildlife Service determines that the proposed action will jeopardize the species, they must offer "reasonable and prudent alternatives" about how the proposed action could be modified to avoid jeopardy. It is very rare to withdraw or terminate projects because of jeopardy to a listed species

Take—The ESA makes it unlawful for a person to take a listed animal without a permit. Take is defined as "to harass, harm, pursue, hunt, shoot, wound, kill, trap, capture, or collect or attempt to engage in any such conduct." Through regulations, the term "harm" is defined as "an act which actually kills or injures wildlife. Such an act may include significant habitat modification or degradation where it actually kills or injures wildlife by significantly impairing essential behavioral patterns, including breeding, feeding, or sheltering." River drying that strands fish or kills trees that ESA-listed birds use, could be considered take under the ESA.

Incidental Take Statement—Through Section 7 consultation, the U.S. Fish and Wildlife Service provides federal agencies with an incidental take statement that identifies the amount of take reasonably expected to occur due to the proposed action. This amount of take is considered lawful provided agencies comply with reasonable and prudent measures (determined by the service and issued in the biological opinion) to minimize take.

Opportunities for Long-Term Bosque Preservation in the San Acacia Reach

Gina Dello Russo, *Bosque del Apache National Wildlife Refuge*

The Rio Grande through New Mexico has been an oasis to settlers for many centuries. It has also drawn wildlife to its abundance of food and shelter within a harsh desert environment. We hope it will survive to do both long into the future. The river is now at a cross roads. Will it survive continuing urban growth along its banks? Will it survive changes in climate? Will the humans presently living along its course recognize the importance of the river in replenishing ground water supplies, assuring wildlife abundance through bosque preservation, and contributing to human health and recreation?

The term bosque as used here includes not only the wooded areas adjacent to the river but all of the plants and animals that live along and in the river, and the dynamics of the river itself. One doesn't exist without the other. Bosque preservation ultimately will include changing the current dense strip of trees, both native and invasive, to a patchwork or mosaic of grasslands, wetlands, shrublands, and forests that we see only in glimpses nowadays. Preservation will require maintaining some of the river processes that will help us design and maintain this mosaic. And finally, preservation will require coordination to ensure that our work is both economically and ecologically efficient.

WHAT ARE THE BENEFITS OF LONG-TERM BOSQUE PRESERVATION?

The benefits of bosque preservation include the creation and maintenance of a habitat mosaic with periodic floodplain flooding. They include reduced fire danger in wildlife habitat and adjacent to private residences. This is accomplished through increased decomposition following flooding, by the active removal of dense invasive forests, and with the maintenance of strategic fire breaks.

It has been shown that the removal of dense stands of salt cedar, followed by the establishment and maintenance of the native grasslands and open forests, results in a reduction of water use by the natural system. The bosque's more efficient use of water makes it easier to maintain the connectivity between open water and shallow ground water, and thus to maintain important habitat areas for endangered species and to reduce losses through evapotranspiration.

Removal of dense salt cedar as part of bosque preservation also assists with water and sediment management. High flows that can spread across a wide floodplain are less likely to endanger the flood-control levee on the west side of the river. Sediment deposition across this wider floodplain during flooding also slows down channel aggradation. Sediment deposition and movement is very important to the plants and animals along this sand-bed river system. These sediments carry nutrients and lay down the seed bed for the establishment of native plants on the floodplain. River flows that are high enough to scour vegetation off sand bars keep the river channel open, increasing the safe channel capacity through the San Acacia reach.

Land values increase when dense stands of invasive trees are removed from private property. A feasibility study completed in 2001 showed that lands with native forests and grasslands were three times as valuable as those with dense salt cedar. Many landowners have voiced their preference for improved native wildlife habitat on their properties.

BOSQUE PRESERVATION IN THE SAN ACACIA REACH TODAY

The communities in the San Acacia reach are making progress toward long-term bosque preservation. The Save Our Bosque Task Force, a diverse group of federal, state, and local government agencies, private landowners, and concerned citizens, has been working together since the early 1990s on issues of public use and recreation, floodplain encroachment, improved biological diversity, endangered species habitat improvement, water use by the natural system, wildfire danger, and invasive species control. Most important, they've done this with the involvement of private landowners and entities from the local area. They have evaluated the potential for bosque preservation in this reach through a feasibility study, planning efforts, and work on a reach-wide monitoring and adaptive management program. These programs provide important information about changes to the river and floodplain. Partnerships with local and regional universities are strong, with much of the research occurring on the Bosque del Apache and Sevilleta National Wildlife Refuges. This research is improving our understanding of bosque water use, fuels reduction, salt cedar control, and habitat restoration techniques.

Adaptive Management

Farmers, developers, industries, other states, salt cedar thickets, the silvery minnow, and even the relentless New Mexico sun all demand their share of Rio Grande water. How should we "manage" these demands so that each gets the appropriate amount? Who decides this, and on what basis? What happens when things change? To date, Rio Grande management has been criticized for not learning from mistakes, for not recognizing research results and new technologies, and for its inflexibility toward new approaches. It has been criticized for not adapting to changing environmental conditions and social needs. Adaptive management can address these issues and provide a structured process that integrates science and allows the flexibility to explore new options, avoid gridlock, and collectively move forward to solutions.

Adaptive management can be defined as an integrated, multidisciplinary approach for confronting uncertainty. It is a philosophy that is used when developing a plan to address environmental and ecological issues. It states that a roadmap should be developed for how to manage, for example, all uses of water in the Rio Grande. All stakeholders should be involved, actions should be taken based on the best science and information currently available, research should be conducted to evaluate success and explore new options, changes should be made to the plan that accommodate the new research, and implementation should continue. Then the cycle is repeated as an open-ended process. It is adaptive because it acknowledges that managed resources will always change as a result of human intervention, surprises are inevitable, and new uncertainties will emerge. It requires that we look at problems in holistic ways and work toward long-term solutions. Adaptive management states that decisions and policies are not merely ends, but means to probe alternatives and understanding in anticipation of future changes and unexpected outcomes. Middle Rio Grande stakeholders are beginning to explore, contemplate, and develop adaptive management strategies on this river system to the benefit of the environment, water managers, and communities.

For more information, visit the Collaborative Adaptive Management Network at www.adaptivemanagement.net.

The wildlife habitats of the bosque in the San Acacia reach are more diverse than in other reaches of the Rio Grande, and they remain relatively healthy, even in these times of limited water supply and changes in water management. Why? Because this reach retains critical physical processes, including occasional floodplain flooding, sediment movement and deposition, and ground water connectivity. The connectivity between the river and floodplain is greatest in this reach of the Middle Rio Grande in terms of continuous river miles. Moderate discharges on the river (3,000 to 7,000 cfs) simulate the flood pulses that scour sand bars, keeping the river channel open, establishing new vegetation on the floodplain, and

The Rio Grande in flood in 1979, looking east across the floodplain at the Bosque del Apache National Wildlife Refuge. Occasional flooding on the river maintains healthy cottonwoods and willows, recharges the shallow ground water aquifer in the valley, and removes vegetation from the river channel so that high water passes safely.

providing diverse aquatic habitat. Efforts are underway to quantify the ecological benefits of flood pulses in terms of flood control, water delivery, and habitat diversity, based on the high spring flows of 2005. However, if flooding occurs at the wrong time of year, it supports the spread of invasive species and promotes increased water use. Summer floods are inevitable, for they come from monsoonal rains on large tributary watersheds or from localized heavy rains. If a reach of river is dominated by flashy summer floods, it will favor salt cedar; if it has both occasional spring floods that establish native plants and flashy summer floods, the native plants will have the edge.

Programs such as the Save Our Bosque Task Force Habitat Restoration Program, the Socorro Soil and Water Conservation District's Invasive Species Control Program, and the Socorro County Wildfire Protection

Program are well established and provide habitat enhancement assistance. These programs also offer technical resources to landowners and managers for continued maintenance of enhanced bosque areas. Since 1999 the Save Our Bosque Task Force has hosted a number of informal meetings for private landowners, government agency staff, and interested citizens, providing opportunities to understand and discuss important bosque issues.

HOW CAN WE CONTINUE TO WORK TOWARD BOSQUE PRESERVATION?

If we are to achieve balance among water users and retain the natural beauty, diversity, and benefits of the river, we need to devote our attention to three broad areas:

Improved Information

- We need to develop models that look at the ecological costs and benefits to changes on the river, including water availability and management, infrastructure, changes in ground water levels, and sediment movement. Modeling efforts from the water management agencies have improved our ability to predict water delivery and movement under different river flow and diversion scenarios. A ground water model is being developed for the San Acacia reach and other reaches of the Middle Rio Grande. But no modeling currently exists to determine the changes to plants along the river as a result of changes in water management or availability. Changes to plants in turn affect the wildlife populations that depend on those plants. Ecological models of the Middle Rio Grande including the San Acacia reach are being developed to improve what we know about the flexibility of the plants and animals along this river. They are needed to answer questions such as: How much drying can occur before stresses result in a die back of existing cottonwoods, willows, and wetlands? What ground water connections are required to sustain a healthy, diverse bosque? What magnitude and frequency of flows are required to maintain the bosque we have today, and to enhance and sustain the bosque into the future? These models will help us understand how much bosque preservation is provided by river flows and how much we, the stakeholders, will need to manage.

- We need to implement monitoring programs that track improvements, and we need to focus on research that will address improved techniques, knowledge, and cost-efficiency of efforts on the ground.

- We need to identify opportunities that will support and increase the river's connection to its floodplain.

- We need to identify opportunities that will provide occasional flood pulses onto that floodplain. Our decision-making ability is limited without tools to predict the plant and animal responses to changing river flow patterns. And long-term sustainability of the bosque ecosystem will require a thorough evaluation of how improved river function and bosque preservation can benefit water management in the future.

Screwbean mesquite grassland at the Bosque del Apache National Wildlife Refuge. These open areas benefit wildlife such as deer and turkey and have high soil salinity levels and dense clay lenses. Such areas would have been found all along the river in the past. If restored, these areas will serve as natural fire breaks on the floodplain.

ACTIVE RESOURCE MANAGEMENT

Active resource management on the Middle Rio Grande would couple water management programs to the other resource programs along the Middle Rio Grande more closely so that, where possible, water management is benefiting other resources. These other resources include healthy riparian areas on the active floodplain, an open floodplain to safely carry high flows without endangering structures, balanced availability of water for recreation on all reaches of the Middle Rio Grande, and infrastructure that provides protection to the valley but allows for long-term sustainability of these other resources. Active resource management requires that water management agencies communicate more effectively with other resource managers and land owners. Specifically:

- We need to support programs that offer landowners alternatives to building houses in flood-prone areas of valley, especially east of the river where no flood control levee exists.

- Where possible, we need to allow the river to occupy more of its floodplain. The current infrastructure of the San Acacia reach was completed in the 1950s. The flood control levees in the San Acacia reach are constructed of unconsolidated spoil material. In many parts of the reach, it is difficult to keep the river flowing because of the ground water gradient to lower-lying lands, the Low Flow Conveyance Channel, and riverside drains. Invasive species are choking the river channel and making it difficult for the river to carry moderate to high flows. Now would be a good time to look at infrastructure changes and improvements that would benefit both economic and ecological aspects of the system for the long term. The federal government is looking at opportunities to widen the floodplain to allow the river more room to work and to provide levee protection. Two areas being considered for improved river/floodplain connectivity are in the Bosque del Apache National Wildlife Refuge and the Tiffany basin.

- We need to support programs that provide landowners with incentives to maintain habitat areas free of fuel buildup and invasive species. The Save Our Bosque Task Force and other interested parties have been developing a program of voluntary conservation easements that, coupled with habitat enhancement on private lands, provide protection from floodplain encroachment, improve biological diversity, and increase land values. This allows periodic high flows for the benefit of water delivery and the bosque.

- We need to provide venues where landowners and land managers can communicate their concerns and ideas for protecting their lands.

Restored wetland feature at the Bosque del Apache National Wildlife Refuge. Constructed ditches, drains, and sleughs deliver water to managed wetlands but also benefit waterfowl and other wading birds. Dense willow stands adjacent to these areas add to wildlife habitat diversity.

IMPROVED COORDINATION

Many different agencies and organizations are involved in efforts toward long-term bosque preservation in the San Acacia reach. And we are making progress here. We need to coordinate those efforts more closely to maximize their effectiveness. Although the specific goals of these different efforts vary, there are common threads: controlling invasive plants; improving forest, scrubland, and grassland health; providing for wildlife and human use; and decreasing the local fire danger. We need to work together more closely on our efforts at habitat enhancement, which include monitoring programs, the timing of on-the-ground projects, and sharing the lessons learned.

Many funding sources are available for certain aspects of bosque preservation, but often these funds go unused. There may be restrictions on how those funds can be used, and annual funding cycles are not always productive. But often the lack of coordination between those working toward habitat enhancement is an issue, as well. Effective salt cedar control and native plant establishment can take a number of years. Those working on bosque enhancement have certainly become well-versed in the different requirements for funding sources, matches, schedules, and restrictions. Coordinated funding sources that allow for comprehensive project implementation would really improve our chances at long-term bosque preservation. We need to review the efficacy of projects funded under existing funding sources, and identify what is missing from these funding sources that would allow for successful completion and maintenance of ongoing habitat enhancement projects.

Some other specific areas where we need to improve coordination of our efforts include:

- Developing programs that provide long-term resources for necessary maintenance of improved habitat areas

- Developing agency partnerships to address key issues:

    –Controlling and preventing infestations of invasive weeds

    –Offering incentives for land managers and owners to participate in programs that protect and benefit the bosque

    –Increasing awareness of the value of the bosque

    –Building partnerships that are committed to long-term bosque preservation

    –Monitoring our progress toward greater bio logical diversity, floodplain protection, fire protection, invasive species control, efficient use of water for the natural system, and endangered species habitat improvement

We must balance river water use with the other water uses in the valley along the Middle Rio Grande. Current local and regional planning efforts must look at what potential water use would be needed to sustain a healthy bosque into the future, and at ways to conserve water for other uses. With a better understanding of the river's water needs, people along the river will be able to make informed decisions about preserving this part of the community. Any program for promoting water balance through management will have to provide venues where landowners and land managers can communicate their concerns and ideas and work toward solutions. I am hopeful that we can make informed decisions that will allow our bosque to continue to thrive, nourishing future generations of people and wildlife.

How Science Can Provide Pathways to Solutions—The Technical Toolbox

Susan Kelly, *Utton Transboundary Resources Center, University of New Mexico School of Law*
Geoff Klise, *Water Resource Specialist*

Policy makers and decision makers often rely upon scientists to provide answers to some of the most pressing problems they face. Scientists have a number of tools at their disposal to do this, including the whole array of technical tools we call "models." Models are generally complex computer programs that deal with real data in an effort to simulate the behavior of natural systems, taking into account an array of variables, from basic physical data—the length and shape of a stream bed, for instance—to complex and often unpredictable variables like rainfall, climate, and future water use. All of this is done in an effort to predict and gage the hypothetical effects of various scenarios so we can understand the impacts of the decisions we make, and chart a course for a future we wish to see.

To address water supply issues in the San Acacia reach, hydrologists and water resource planners use regional models that address the broader Upper Rio Grande watershed in combination with models that look specifically at hydrology in the San Acacia reach. These dynamic modeling tools help us understand the workings of the natural hydrologic system, the riparian ecosystem, and the human impacts on water supply in this region. Understanding these complex relationships, and trying to predict how they will interact in the future, is the objective of modeling. This paper offers a look at some of the more significant efforts at developing and using hydrologic models and other technical tools applicable to the San Acacia reach.

HYDROLOGIC MODELING—THE UPPER RIO GRANDE WATER OPERATIONS MODEL

The basic tool for water supply planning in the Middle Rio Grande is the Upper Rio Grande Water Operations Model, or URGWOM. This is a multi-agency water operations model designed to help manage the accounting and operational decision making of many agencies. It is currently used on the Rio Grande throughout New Mexico to simulate water storage and delivery operations, to model flood control operations, and to provide a basis for long-range planning from thousands of pieces of information on water use, climate, evaporative losses at reservoirs, see page to

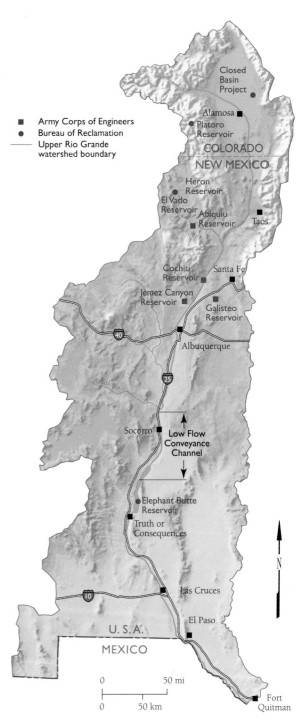

Map of the Upper Rio Grande Basin, the area where water operations can be simulated in the Upper Rio Grande Water Operations Model (URGWOM).

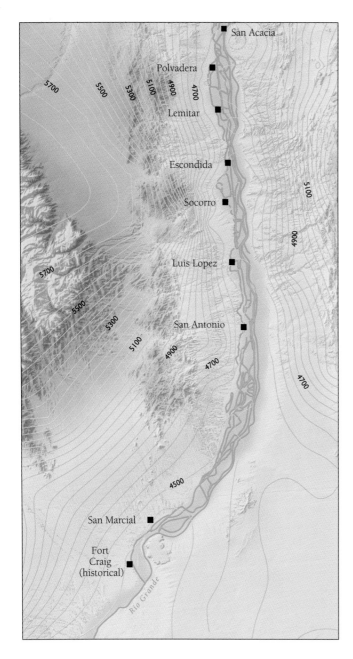

Map of simulated water table, using linked surface water and ground water model for Socorro and San Marcial Basins between San Acacia and Elephant Butte Reservoir. To understand the general water movement in the shallow aquifer, monitoring wells in the study area were used to develop a water table map. In general, ground water moves from east to west to the center of the basin, where it discharges to the surface water. The water table map also indicates a strong north-south hydraulic gradient.

ground water, snowmelt runoff, and other hydrologic variables. It provided forty years of flow and storage projections for the Upper Rio Grande Water Operations Review and Draft Environmental Impact Statement, which evaluated alternatives for future water management.

URGWOM was developed by the U.S. Army Corps of Engineers and U.S. Bureau of Reclamation, with significant participation by the U.S. Geological Survey and the state of New Mexico. It continues to be refined and improved, and our ability to use it for planning purposes in conjunction with other modeling tools is expanding. To model actions involving changes in how federal agencies operate the reservoirs on the Rio Grande and requiring changes in legislative authority, we would have to revise our current model. The Interstate Stream Commission is leading work by the URGWOM tech team that will allow the model to simulate the interaction between the shallow and the deep water aquifers. This revision will allow the model to depict more accurately how water is routed between Cochiti and Elephant Butte.

THE MIDDLE RIO GRAND WATER SUPPLY STUDY

The Middle Rio Grande Water Supply Study, Phase 3, conducted by S.S. Papadopulos & Associates in 2004, evaluated the regional water supply. The conjunctive ground water–surface water supply available to the Middle Rio Grande region, under the constraints of the Rio Grande Compact, is characterized under a range of conditions. The study evaluates the probability of compliance with the Rio Grande Compact, assuming projected demand through year 2040. The study relied on demand projections as developed in the Jemez y Sangre, Middle Rio Grande, and Socorro–Sierra regional water plans, and assumes implementation of management actions suggested by the plans. The study concluded that the Middle Rio Grande region would likely have a severe water deficit in 2040 without implementing the water plans, and that even with full implementation (a highly optimistic future scenario) there would remain a projected deficit.

SAN ACACIA SURFACE WATER/GROUND WATER MODEL

The San Acacia Surface Water/Ground Water Model was created to improve our understanding of the complex interactions between the surface and subsurface hydrologic systems in the Socorro and San Marcial basins. Developed by the Interstate Stream Commission, the model simulates the Rio Grande channel, the Low Flow Conveyance Channel (LFCC), the main irrigation canals and drains, and the alluvial and Santa Fe Group aquifers in the reach from San

Acacia to Elephant Butte. The purpose of the model is to evaluate potential system-wide depletions that may result from various actions, including operation of the LFCC, implementation of habitat restoration projects, and modifications (both natural and man-made) of the river channel. A recent update uses a high-resolution telescopic model that focuses on the riparian area from the river west to the LFCC to predict the effects of habitat restoration between Highway 380 and San Acacia on water supply.

COOPERATIVE MODELING IN THE MIDDLE RIO GRANDE

In the late 1990s regional water-planning efforts in the Middle Rio Grande were initiated by the Middle Rio Grande Water Assembly and the Mid Region Council of Governments. Sandia National Laboratories was contracted to develop a decision support tool to conceptualize how water is used in the region, to understand the complexities of the system, and to recognize tradeoffs and consequences with different conservation approaches. Although the planning was done above the San Acacia reach, the model looks at flows into Elephant Butte for the purposes of meeting the obligations of the Rio Grande Compact with Texas. The model represents complex interactions and feedback between physical and social systems. Sandia National Lab included components such as surface water, ground water, population, and demands from urban use, agriculture, evaporation, and environmental uses.

The model was created in a collaborative fashion, with members of the planning groups giving input to the modeling team. The Utton Center at the University of New Mexico provided facilitation to the model development team, acting as an impartial party to manage the meetings and foster communication. Because of the number and variety of participants, facilitation was needed to organize input and bring closure to the discussion of issues. The simulation results gave the planning group a preferred scenario that was used as the platform for finishing the regional water plan. Local governments in the region, including the cities of Albuquerque, Rio Rancho, and other municipalities, adopted the plan as the "guidance document" for their own planning efforts. Implementation of the plan, as with the Socorro–Sierra and the other regional plans mandated by the state of New Mexico, will require significant action on the part of many entities in the region.

Building on the success of the Middle Rio Grande

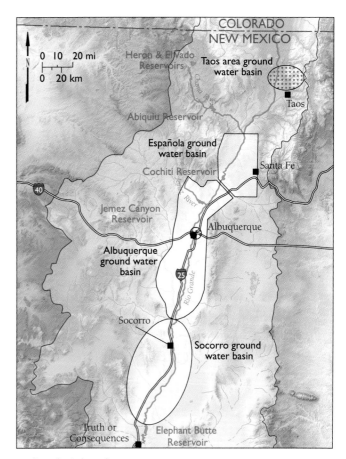

Integrated Hydrologic Modeling, an interactive planning tool for the Upper Rio Grande.

Water Assembly collaboration, Sandia National Laboratory is currently working with the URGWOM team to develop a complementary system dynamics model, based on the original Middle Rio Grande planning model. Integration of both models will improve management of water resources in the Rio Grande basin because of the ability to model the interaction between surface and ground water. Decision makers can use the model to understand the impacts of reservoir operations on the river and ground water systems.

In the San Acacia reach, there are many different modeling efforts underway that ultimately will be included in a unified water operations model for the Rio Grande. Watershed models are being developed by New Mexico Tech; the Rio Salado will be the first. A mortality model for the Rio Grande silvery minnow was built to try and understand how water quality affects

What Is Modeling?

In general terms, a model is a simplified representation of a complex real system. Because it is very expensive and time consuming to test the effects of management changes on a real hydrologic system, we take a shortcut and develop a model of each aspect of the system that we need to understand. Each model must be complex enough to include all the phenomena and structures that are important to us, but not so complex as to be mathematically insolvable.

The structure of a model is developed using basic information about the system we are simulating—for example, the length and width of the streambed for surface water models, and the nature of the rocks that make up the aquifer system for ground water models. The system is divided up into grid cells or nodes, each of which represents a small chunk of the system.

Input to a ground water or surface water model includes the inflow of water (aquifer recharge in the case of a ground water model, and flow from upstream and from tributaries to a surface water model), as well as diversion of water from the system. A model uses basic equations that govern the flow and conservation of water (like Darcy's Law) to keep track of this water and move it along at the proper velocity, from cell to cell or node to node, and determine its fate.

A ground water model calculates what the water levels in the aquifers will be, and how much ground water will discharge into adjacent streams A surface water model calculates how much river water makes it downstream, how fast it gets there, and in the case of complex, rule-based models, how much is diverted from reservoirs for irrigation, how much is released from the reservoir into the stream bed, and how much remains in reservoir storage.

—Excerpted from an article by Peggy Barroll et al. on hydraulic modeling, which appeared in our 2003 Decision-Makers Field Guide, *Water Resources of the Lower Pecos Region, New Mexico*.

silvery minnow populations. In addition, data from the Interstate Stream Commission's San Acacia Surface Water/Ground Water Model will also be included. Because actions taken upstream affect water flows in the San Acacia reach, integrating water operations with Sandia's model will provide decision makers with a comprehensive set of tools that can help decipher the relationships between physical and social systems on the river between San Acacia and Caballo Reservoir.

SOCORRO–SIERRA REGIONAL WATER PLAN

The Socorro Soil and Water Conservation District (SWCD) is a government subdivision of the State of New Mexico very active at helping to direct funding and education from a variety of sources to those at the local level within the San Acacia reach. The SWCD was designated as the fiscal agent for writing the Socorro–Sierra Regional Water Plan. (The San Acacia reach lies within this planning region.) The Interstate Stream Commission accepted the regional plan in 2004 as the guiding document for planning efforts in Socorro and Sierra Counties. Prepared by Daniel B. Stephens & Associates, in cooperation with a wide variety of professionals and interested citizens, the plan contains a wealth of information on the region's projected demand for water, assesses ground water and surface water supplies, and evaluates alternative future scenarios for balancing supply and demand. There is still much work to be done to implement the three regional plans between Otowi and Elephant Butte and to reconcile their recommendations. There are conflicts among them, particularly concerning the transfer of water rights from agriculture to urban uses.

EVAPOTRANSPIRATION MODELING

Evapotranspiration is one of the most significant depletions on the river; therefore, tools to model evapotranspiration are critical. The ET Toolbox, a modeling tool developed by the U.S. Bureau of Reclamation, is the evapotranspiration model for the Middle Rio Grande. The primary purpose of the ET Toolbox is to estimate daily rainfall and water depletions (both agricultural and riparian) and open water evaporation within specific reaches. For operational and management purposes, the ET Toolbox provides products by river reach and by Middle Rio Grande Conservancy District (MRGCD) division to show various consumptive use requirements. These daily values can be used by URGWOM.

A high density evapotranspiration network, with real-time state-of-the-art instrumentation and modeling integrated with real-time remote imagery, is being developed as a collaborative project between the University of New Mexico, New Mexico State University, and New Mexico Tech in a project known as EPSCoR. The hydrology component of EPSCoR will result in coupling and extending models for climate and hydrologic predictions and increasing the connectivity between ground-based and satellite-based data. The main objective is to extend and integrate a network of telemetered instruments that provide ground-based measurements of evapotranspiration in different ecosystems (riparian, upland, and agricultural). The primary product will be high frequency, high resolution evapotranspiration maps for the Rio Grande watershed between Cochiti Reservoir and the Mesilla Valley. Data products will be prepared and distributed via the Internet in a form accessible to researchers, managers, and water users.

DECISION SUPPORT SYSTEM FOR THE SOCORRO DIVISION

The Middle Rio Grande Endangered Species Act Collaborative Program is a multi-agency group of stakeholders trying to address in a cooperative manner science, habitat, and water supply issues for endangered species. Together with the Interstate Stream Commission they funded development of an effective rotational water delivery system for the Belen Division of the MRGCD in FY 2003. In FY 2004 the decision support tool was extended to the Socorro Division. Through work accomplished by Colorado State University and S. S. Papadopulos & Associates, a scheduled, rotational water delivery system for irrigators was designed. This replaced the continuous, on-demand delivery of the past. The rotational delivery system has resulted in significant reductions in water diversion in the Belen and Socorro Divisions of the MRGCD. There is a need to improve the model and its data sets, including validation of assumed values of irrigation efficiency, soil moisture depletion, and the extent of conveyance losses in delivery channels. The assumptions need to be compared to field conditions, and the return flow functions need improvement.

The MRGCD has actively embraced the rotational delivery system. The new operational system, combined with other improvements such as new flow meters on all diversions and delivery canals, automated water control gates on diversion dams and canals, limited lining of canals, and other improvements to the water conveyance system, has allowed the MRGCD to reduce depletions by 47 percent since 1996. In spite of severe drought conditions that have drastically reduced upstream storage of water during this time period, MRGCD has been able to provide an adequate supply of water to farmers throughout the district. This has also helped the MRGCD and other water managers keep enough water in the Rio Grande to protect the endangered silvery minnow and southwestern willow flycatcher and thus avoid threatened legal impediments.

GEOGRAPHIC INFORMATION SYSTEM AND REMOTE SENSING DATA

Operating on the University of New Mexico campus in Albuquerque, the Earth Data Analysis Center (EDAC) houses an extensive repository of Geographic Information System (GIS) data for New Mexico. The data can be downloaded by anyone and viewed with either standard commercial software or open source software. The clearinghouse includes data on general boundaries, roads, cities, topography, climate, geology, soils, elevation, water resources, aerial photographs, remote sensing, and population.

Because these data cover the entire state of New Mexico, specific information regarding the San Acacia reach can be extracted. Besides the statewide coverage, there are some datasets that are specific to the San Acacia reach, such as detailed land use inventories, trends, and vegetation maps. These data are constantly updated and represent the best collection of publicly available GIS data that can be used by decision makers in the San Acacia reach.

In 2003 NASA funded a center at the University of New Mexico devoted to acquiring real-time remote sensing data. The Center for Rapid Environmental Assessment and Terrain Evaluation (CREATE) uses existing satellites and a new ground station to acquire and process data in a very short amount of time. In addition, the data are available at much higher resolution than other remote sensing products. These data can be used in decision support systems to help understand evapotranspiration rates, for snowpack analysis, fire condition, vegetation growth, and landscape changes. In the San Acacia reach, the CREATE group has been supporting the evapotranspiration work being conducted as part of the EPSCoR program.

ECOSYSTEM RESEARCH

Located in the San Acacia reach, the Sevilleta National Wildlife Refuge's Long Term Ecological Research Center (LTER) has been conducting research on how climate change can impact ecosystems. Sevilleta is unique in that many biotic zones intersect within the refuge; the area can thus provide sensitive indicators for environmental response to climate change. Recent research projects have looked at the impact of climate change on vegetation, the response of vegetation to prescribed burns and cattle grazing, and evaporation and transpiration in uplands and riparian areas near the Rio Grande.

Sevilleta houses a large "spatial database" for both

GIS and remote sensing data that have been collected as part of the research conducted at the LTER. Some of the data are from publicly available sites and clipped into the LTER boundaries. All of the research conducted at Sevilleta is within the San Acacia reach, although some projects have locations both inside and outside Sevilleta, such as the evapotranspiration research conducted by EPSCoR.

The San Acacia reach has many sophisticated tools and models available to provide guidance on our water future: water for future human use, water for agriculture, water to meet the terms of the Rio Grande Compact and a treaty with Mexico, and water for the ecosystem. Models and studies are continually being refined, updated, and improved as better data are obtained—but the basic tools are in place. The extreme variability of water supply on a year-to-year basis, and the inherent difficulty in predicting variability in the future, make "certainty" a challenging target.

Models and planning tools can provide information to help policy makers make the most informed decisions. The hardest work remains in the arena of public policy. Obtaining the legal mechanisms to administer water rights above San Acacia is essential to achieving long-term sustainability in the San Acacia reach. It is critical that the three regional plans in the areas that affect the water supply in the San Acacia reach be consistent in both recommendations and implementation. Allowing scientists to use the models and propose options to managers and policy makers without being unduly constrained by political issues is the best hope for arriving at sustainable solutions.

Suggested Reading

For a hands-on look at one hydrologic model, the Sandia National Laboratories Middle Rio Grande Cooperative Model (2005), go to http://nmh2o.sandia.gov/ExTrainSD/SDWelcome.asp

The Socorro Soil and Water Conservation District Regional Water Plan is available at http://www.socorroswcd.com Click on the link that says Regional Water Plan for information on projected water supply and demand in the region.

For more information about the basin and the San Acacia reach hydrology, visit the U.S. Army Corps of Engineers at http://www.spa.usace.army.mil/urgwom

For detailed information about the Middle Rio Grande water supply and demand issues, have a look at the S. S. Papadopulos & Associates' Middle Rio Grande Water Supply Study, Phase 3, available at http://www.ose.state.nm.us/isc_planning_mrgwss.html

CHAPTER FOUR

A VISION OF THE FUTURE

**DECISION-MAKERS
FIELD CONFERENCE 2007
San Acacia to Elephant Butte**

The Rio Grande above San Marcial

What We Stand to Lose in the San Acacia Floodway

Rolf Schmidt-Petersen, *Rio Grande Bureau, New Mexico Interstate Stream Commission*
Peggy S. Johnson, *New Mexico Bureau of Geology and Mineral Resources*

Water resource issues in the Middle Rio Grande are complex and interlinked, and nowhere more so than in the San Acacia reach. It is difficult to develop solutions to comprehensively address the management of interconnected water, land, and ecosystem resources within the middle valley, given the differing expert viewpoints on resource needs and priorities, the different physical characteristics of each reach, and possible unintended consequences of particular solutions. Many decisions have been driven by a specific event or regulatory mandate and are generally aimed at the most dire pending disaster. Management of river flows in the Middle Rio Grande valley entails a balance of competing interests and safety issues. In the face of potentially disastrous high river flows, we have to act immediately to prevent breach of a levee and protect humans, infrastructure, and the environment. On the other hand, how can we better manage high flows to provide benefits to the environment, the river channel, and the long-term functionality of the river system? How does a decision maker prioritize competing interests and decide what to support and what not to support? The cost of extended inaction in the San Acacia reach is loss of the agricultural, environmental, and economic productivity of the San Acacia reach as well as much of the Middle Rio Grande. In this paper we describe one successful outcome from recent work conducted in the San Acacia reach and illustrate a possible worst-case scenario and its potential impacts.

CHALLENGES AND OPPORTUNITIES OF THE 2005 FLOW EVENT

During periods of low river flow, the U.S. Bureau of Reclamation, the New Mexico Interstate Stream Commission, and other agencies coordinate on floodway projects that reduce flood risk, aid in routing water safely through the middle valley, and control natural water losses. Currently these projects include construction and recurring maintenance of the pilot channel, selective reinforcement of river levees, maintenance of drains, and removal of sediment plugs from the river. During high river flows, the Bureau of Reclamation and the U.S. Army Corps of Engineers implement their emergency response capabilities, respectively, by shoring up levees and implementing other short-term fixes and initiating flood-control operations from upstream reservoirs. Over the past five years, the average cost of such work within the Middle Rio Grande floodway has exceeded $10 million per year in state and federal funds. Although these efforts have reduced flood risk, the flood threat to adjacent lands remains throughout the Middle Rio Grande floodway, including the San Acacia reach. There are historical, naturally flood-prone areas, and both engineering and natural events affect flood potential in a number of ways. These include channel narrowing, increased vegetation density and encroachment into the river channel, aggradation of river sediment, disconnection of tributary arroyos, deteriorated levees and non-engineered levees, and infrastructure in the floodplain such as residential development and the San Marcial railroad bridge. These areas constrain the rate of safe releases from upstream reservoirs.

During spring 2005 the Middle Rio Grande basin experienced the highest snowmelt runoff in about 10 years. To accommodate the high flow, the Corps of Engineers attempted to reach the maximum authorized safe-channel-capacity release from Cochiti and Jemez Canyon Reservoirs. The corps was able to maintain a high release for an extended period of time, primarily due to the preventative and emergency repair work by the Bureau of Reclamation during the runoff, and successfully managed potentially disastrous flood flows, accruing a number of benefits to water users, endangered species, and the ecosystem.

The successes of 2005 were many. Through skillful management of flood flows and a small dose of serendipity, the corps was able to maintain the high consistent flood release for several weeks without a levee breach. As a result, significant overbank flooding occurred within the Middle Rio Grande bosque, and the Article VII storage prohibition of the Rio Grande Compact was lifted on the day of peak runoff into El Vado Reservoir. Lifting of the storage prohibition allowed the Middle Rio Grande Conservancy District (MRGCD) to store more than 120,000 acre-feet of water in El Vado Reservoir with about 80,000 acre-feet of that storage occurring in the ten days immediately following the peak. The storage in El Vado Reservoir provided water to farmers and helped sustain river flows for the silvery minnow during fall 2005 and 2006. However, if the 2005 flood releases had been further restricted or a levee had been breached anywhere in the valley, a very different scenario might have played out: The storage prohibition would not have been lifted in May, MRGCD would not have amassed adequate storage in El Vado Reservoir, significantly reducing water deliveries for farmers and the minnow, and flooding may have occurred outside the levee system in vulnerable areas of the Middle Rio Grande Valley.

Erosion of river bank at river mile 111, just a few miles below San Acacia, due to high snowmelt runoff in spring 2005. The river is migrating laterally to the left, toward the cut bank.

The maximum flood release in 2005 produced a flow of about 6,000 cfs at San Acacia, providing significant benefits to the river channel and riparian habitat in some areas, but increasing flood risk in others. Overbank flow inundated some areas of cottonwood bosque that hadn't been flooded in several years, rejuvenating the riparian system and the Rio Grande silvery minnow. Adult minnow catches in the Middle Rio Grande in fall 2005 were some of the highest on record. Where it occurred, scouring of the main channel stripped congested vegetation from sand bars, thus helping to maintain an open channel. While scouring increased flow capacity in some areas, it also increased flood risk where erosion allowed the river to migrate laterally toward a levee. The Low Flow Conveyance Channel and adjacent levee have been moved back from the river in one location south of San Acacia to reduce flood risk by accommodating greater migration of the channel. Sediment scoured by high flows in upstream reaches was deposited in downstream reaches, plugging the river channel for almost two miles near Tiffany, four miles south of Bosque del Apache National Wildlife Refuge, and severely straining the adjacent levee. Emergency measures were conducted during the high flows, and the sediment plug was excavated in the fall of 2005. The unfortunate reality is that the Corps of Engineers cannot release sufficient water to cause inundation of the bosque in the Cochiti and Albuquerque reaches without increasing the flood risk through Isleta and San Acacia.

The 2005 water season also produced benefits to downstream users. Due to the high inflow to Elephant Butte Reservoir during the spring of 2005, the water elevation in the reservoir rose sharply, and recreation interests experienced a relatively good year. The Bureau of Reclamation allocated a nearly full supply of surface water for the Elephant Butte Irrigation District, the El Paso County Improvement District No.1, and the Republic of Mexico.

A WORST CASE FLOOD SCENARIO FOR THE SAN ACACIA REACH

Despite the successes of 2005, another flood scenario threatens the Middle Rio Grande valley—one with a less beneficial outcome—if we fail to comprehensively manage water, sediment, and riparian system function. The Rio Grande is sediment-laden with vast sources of sand, silt, and clay immediately available in its adjacent terrain. A worst-case scenario for the San Acacia reach can result from a combination of sediment load and high flood flows from record snowmelt runoff or intense summer rains similar to those experienced in summer 2006. South of Highway 380, the river is actively building up its channel with sand deposited from its sediment heavy waters, and it will continue to aggrade. In response to a high flow event, we anticipate that the river will again become plugged with sediment near Tiffany, as it did in 2005. What happens next depends on our level of preparation, our ability to respond, and the whim of Mother Nature. If

Aerial view of the sediment plug near Tiffany in spring 2005, when the plug formed. The Low Flow Conveyance Channel is visible on the right; Black Mesa is visible at the top of the photo.

A close up of the sediment plug near Tiffany in fall 2005. The excavation in the center of the photo has revealed the location of the water table, several feet below the surface of the plug.

we are unlucky and the river cannot scour a path through its clogged channel, then the sediment plug will become vegetated, and the river channel will effectively disappear. When this happened in the 1950s, various pilot channels and the Low Flow Conveyance Channel were constructed over a number of years to facilitate movement of water and sediment to Elephant Butte Reservoir. If this scenario were to be repeated today, the human and environmental consequences would be disastrous in comparison.

A flood of the future—As sediment builds up or aggrades the river channel south of Highway 380, possibly becoming congested with dense vegetation, private land and homes upstream on the east side will be vulnerable to flooding at lower river flows. Land and homes on the west side of the Rio Grande have a lower flood risk because of limited protection provided by the spoil-bank levee adjacent to the Low Flow Conveyance Channel. The magnitude of reservoir flood-control releases available will likely decline because of levee integrity issues in the San Acacia reach. Ultimately, a high flow from the uncontrolled watersheds of the Rio Puerco, Rio Salado, or both, possibly in combination with an already high reservoir release flow, will cause a levee failure somewhere south of San Acacia. Once out of its banks, the river will quickly fill the Low Flow Conveyance Channel with sediment, flow from the failure point toward low-lying areas of the valley, and then erode through its existing channel in an upstream direction, while spreading floodwaters into the Socorro valley.

New Mexico's compact deliveries—In part because the river channel ceased to exist near Tiffany in the late 1940s and early 1950s, New Mexico's water deliveries to Elephant Butte Reservoir suffered. New Mexico was out of compact compliance from 1948 through 1968, and Texas filed suit against New Mexico in response. Due to reductions in the quantity of water reaching Elephant Butte Reservoir, usable storage fell below 400,000 acre-feet and Article VII compact restrictions on upstream storage were in place in many years, reducing the surface water available for use by MRGCD and Santa Fe from 1950 until about 1980. Future flood damage to conveyance works in the San Acacia reach would again degrade our ability to convey water to Elephant Butte Reservoir, with potential outcomes not unlike those that occurred in the 1950s, 1960s, and 1970s.

The surface water supply and agriculture—For farmers in the Socorro Division of the MRGCD and at the Bosque del Apache NWR, the immediate impact of a levee failure would depend on the location of the failure. If the failure occurred downstream of Bosque del Apache National Wildlife Refuge, the human and environmental impacts would be minimized. A water table rise would develop in the floodplain from clogging of the Low Flow Conveyance Channel and nearby drains, ruining crops and irrigable lands. If the failure occurred in the main farming area north of the refuge, many low-lying farms would be inundated and filled with sediment or waterlogged. As the MRGCD drainage system became clogged or overloaded or both, affected areas would become unusable. Surface water supplies would be limited for farmers outside of the Socorro valley in the Elephant Butte Irrigation District, Texas, upstream in the MRGCD, and for citizens of Santa Fe.

Endangered species—Both the Rio Grande silvery minnow and the southwestern willow flycatcher would be imperiled by an extreme flood event. The river channel downstream of a levee failure would dry, and fish residing in that reach would die. The river channel upstream of a levee failure would incise, or lower in elevation as water and sediment flowed out into the valley. Some fish would be carried onto the inundated floodplain, and aquatic habitat in the incised reach would be significantly altered. Existing and suitable southwestern willow flycatcher habitat would be impaired if channel changes eliminated nearby moist substrate and open water that provides favorable nest sites. Finally, if upstream storage in El Vado Reservoir is restricted, less water would be available for release during normal periods of low river flow, and the San Acacia reach would be more prone to channel drying.

Bosque and riparian health—The long-term viability of the cottonwood bosque, historic east side wetlands, and other riparian vegetation would also be jeopardized during a worst-case flood event. Where the river channel has aggraded and is perched above the floodplain, a levee failure would cause the river channel upstream to incise, or lower in elevation, and the downstream channel to dry as water and sediment flowed out into the valley. Both channel incision and river drying would lower the water table, and many native trees would die if the water table dropped below the active root zone. In low-lying areas that cannot be reclaimed when floodwaters recede, salt cedar and other invasive species we have been trying to eradicate could move in and overwhelm the flooded areas.

CAN WE PREVENT A WORST-CASE FLOOD SCENARIO?

If a worst-case flood scenario comes to pass, there will be little hope of restoring the agricultural, environmental, and economic productivity of much of the San Acacia reach. It is not certain whether this future flood disaster can be prevented unless some large-scale project is implemented. We do know that the federal and state dollars currently allocated are insufficient to keep up with the number of priority sites in the middle Rio Grande.

The Bureau of Reclamation, the Corps of Engineers, and the Interstate Stream Commission annually request funding to address flood potential in the Middle Rio Grande. Reclamation currently requests approximately $8 million per year for design, permitting, construction, and monitoring of flood-control projects. Except for about $20 million in congressional funding several years ago (used to buy new equipment, relocate the Low Flow Conveyance Channel and levee near San Acacia, and implement a few other high priority projects), the federal budget has been flat. Although the Bureau of Reclamation has used the

An amphibious excavator clearing the Tiffany plug in fall 2005.

funding to construct a number of fixes, the number of priority sites keeps rising. Currently, the Bureau of Reclamation lists 26 priority levee sites in the Middle Rio Grande Project. Examples of priority sites in the San Acacia reach are located at river miles 113 and 114 and the Tiffany area, where projects to relocate or raise the levees and/or realign certain sections of river are planned. Additional priority projects include removal of sediment plugs, widening the river channel, and maintaining the Elephant Butte delta pilot channel.

State and federal partnerships on collaborative projects are essential for managing and preventing flood disasters in the Middle Rio Grande valley. The Corps of Engineers oversees and directs operations of the river and reservoir system during certain high flow events, implements flood fighting activities, and aids state and county emergency managers. The federal agency also has the capacity to design and build engineered levees and various flood-control facilities to protect farmland, homes, and cities. The Interstate

Stream Commission provides a federal cost share for selected levee projects and annual floodway maintenance, focusing its efforts in areas that are outside of federal or MRGCD scope. The agency also excavates the lower 11 to 15 miles of the Elephant Butte pilot channel and conducts projects with the Bureau of Reclamation to maintain river conveyance (like removing the Tiffany sediment plug). The Interstate Stream Commission has spent an average of about $2.8 million per year over the past five years on such activities. The MRGCD is generally responsible for maintaining the levees within its boundaries.

Forging a Sustainable Water Policy in the Middle Rio Grande Valley—a Downstream Perspective

Peggy S. Johnson, *New Mexico Bureau of Geology and Mineral Resources*
Mary Helen Follingstad, *Santa Fe Regional Planning Authority*

The strategy where one jurisdiction tries to achieve water solvency at the expense of a neighbor is generally unpalatable and strictly prohibited in the Interstate Stream Commission's regional water planning guidance. In the final analysis, what does the state of New Mexico do when the Middle Rio Grande region has garnered the entire valley's water yet continues to demand more?

The face and character of the Middle Rio Grande valley have changed over time, as all things do. Most of the changes were not planned; rather, they happened gradually. In the coming decades change will continue to accelerate and threaten the economic base and cultural identity of the valley's rural communities.

New Mexico is a water-limited state and since the 1990s has been one of the fastest growing states in the nation. Explosive population and commercial and economic growth in the Middle Rio Grande valley have been deemed lucrative, progressive, and inevitable and have fueled a market-driven competition between cities and farms for the rights to use the limited waters of the Rio Grande and its adjacent aquifers. The collective demands of growth, environmental and ecologic needs, ever-increasing ground water pumping and stream depletion, system inefficiencies, and the unknown risk of climate disruption threaten to force default on New Mexico's contract to deliver water to downstream users in southern New Mexico and Texas. Consequences of the boom loom large for cities, counties, water utilities, rural and agricultural areas, the environment, and the present distribution of water rights. With limited opportunities to develop or import new supplies, new uses must rely largely on water obtained from changes of existing uses. In the Middle Rio Grande valley, the reallocation of water to "higher valued uses" is accompanied by adverse consequences for rural communities, agriculture, and the environment, all of which must be considered.

WATER POLICY AND PLANNING—MITIGATING NEGATIVE CONSEQUENCES OF GROWTH

As water use moves from farms to cities, it can trigger unforeseen social, economic, and environmental consequences that impact the quality of life for all who live in the Middle Rio Grande valley. Over-appropriation; hydrologic imbalance; the decline of agriculture; environmental, ecological, and recreational impacts of dwindling surface water supplies; and degraded water quality—these are all things that happen when you transfer large blocks of water rights from place to place over long distances and from surface water to ground water. The enormity and interconnection of these issues underscore the inter-relationships of growth, sustainability, and water policy. Although decisions about how or where to grow are rarely influenced by either water policy or availability, there is little question that future growth must consider natural resource constraints.

The realization that both water and our capacity to grow are limited is not new. Neither is the effort to develop plans that promote orderly transitions and alleviate unwanted consequences that are sure results. A comprehensive look at the impacts and options regarding growth by the New Mexico Department of Finance Administration concluded in 1996 that "we are a water limited state…we are all wildly borrowing against the future…[P]erpetual growth…is impossible for New Mexico…why are we afraid of statewide planning?"

The Western Governor's Association, in their 2006 report on *Water Needs and Strategies for a Sustainable Future*, made these pertinent recommendations promoting fairness, balance, and sustainability in growth policies:

- States should identify water requirements needed for future growth, and develop integrated growth and water supply impact scenarios that can be presented to local decision makers.

- States should facilitate collaborative watershed-focused planning that balances desirable growth and protection of the natural environment, which depends on surface and ground water quantity and quality.

- In reviewing applications for new water uses, transfers, and changes in use, including in-stream flows, states should consider local, tribal, and watershed plans and decisions regarding growth management.

- States and local government should consider the impacts of continued growth that relies on transfers from agriculture and rural areas, and identify feasible alternatives to those transfers.

Each recommendation embraced local or watershed-scale collaborative planning and the involvement of state and local government in developing solutions that balance sustainable growth and water use. The first recommendation has generally been accomplished in New Mexico; the remaining three have barely advanced beyond the level of round-table discussion. Although growth management has primarily been a local matter, states have a critical role to play. The state engineer has the primary responsibility for water allocation and management, and the jurisdiction to sanction or restrict the transfers of existing uses that fuel urban growth. The state engineer and the Interstate Stream Commission also have authority and responsibility for initiating and assisting water planning on both a regional and state level.

In 1986 a federal court decision (*City of El Paso ex rel. Pub. Serv. Bd. v. Reynolds*, 563 F. Supp. 379 (D.N.M. 1983)) upheld a challenge from El Paso to a state statute prohibiting export of water to Texas. The decision provided: (1) that an out-of-state applicant proposing to appropriate New Mexico's ground water must prove that all efforts at conservation have been exhausted; and (2) that the New Mexico state engineer could reject an out-of-state water rights application where New Mexico could show that local water was needed to meet projected demands. In response to this decision, the New Mexico legislature in 1986 appointed a team of experts to investigate the status of the state's water resources and its relationship with future demand. The investigation kicked off New Mexico's water planning effort and found, among other things, that:

- Even with extensive efforts toward water conservation, the effects of converting surface water rights from agriculture to municipal and industrial uses would weaken the agricultural economy significantly in a relatively short period of time. With significant (25 percent) water conservation, the Middle Rio Grande was projected to lose 10 percent of its agricultural water rights by 2003, 25 percent by 2033, and half of its agricultural water by 2071.

- Transferring water to areas of the state needing to import water "will create conflict between the source area and the area to which the water is transported."

- "There may be areas of the state that need preservation because the culture or the land or both constitute irreplaceable assets. [I]t is unwise to allow the very best agricultural lands to go out of production. ... Agriculture may not be able to compete with municipalities and other industries for water from a strictly economic viewpoint. Yet, the long-term interest of the state may best be served by sustaining a healthy agricultural industry in selected areas."

- The state should work in partnership with each region to develop a series of regional water plans, compile regional plans into a state water plan, and form state-regional partnerships for water development and cooperation in promoting water conservation.

Twenty years later, the products of this foundation are sixteen regional water plans that quantify water supply and demand and that identify possible regional solutions to rectify supply-demand gaps. In 2003 legislation was passed that authorized the State Water Plan to be "a strategic management tool for the purpose of:

- Promoting stewardship of the state's water resources;

- Protecting and maintaining water rights and their priority status;

- Protecting the diverse customs, culture, environment, and economic stability of the state;

- Protecting both the water supply and water quality;

- Promoting cooperative strategies, based on concern for meeting the basic needs of all New Mexicans;

- Meeting the state's interstate compact obligations;

- Providing a basis for prioritizing infrastructure investment; and

- Providing statewide continuity of policy and management relative to our water resources."

The State Water Plan, completed in 2003, provides guidance and policy on the linkage between land and water use. Policy statements in the State Water Plan mimic the Western Governors' Association recommendations and require "consideration of the relationship between water availability and land-use decisions" and water rights transfer policies that "balance the need to protect the customs, culture, environment, and economic health and stability of the state's diverse communities while providing for timely and efficient transfers between uses." Pursuant to legislation, the State Water Plan is to be updated in 2008.

REGIONAL CONFLICTS AND LAND USE ISSUES IN THE MIDDLE RIO GRANDE VALLEY

Strategies to manage growth, conserve water, develop new water supplies, protect water quality, and protect quality of life have emerged in regional water plans where water resources are insufficient for existing or projected growth or where cultural impacts exist with respect to water transfers. Projects have been identified statewide, but local know-how, funding, and guidance from the state are often lacking, and many regions are uncertain about how to proceed with implementation. One serious obstacle to plan implementation is resolving disparities between plans and conflicts among regions relying on the same water source. Nowhere in the state are the consequences of not implementing solutions more serious than in the Middle Rio Grande valley, and regional water plans provide a road map, albeit a sketchy one.

Three water-planning regions, each with a recently completed water plan, lie along the Rio Grande between Otowi Gage and Elephant Butte Reservoir: Jemez y Sangre (Santa Fe, Los Alamos, and Rio Arriba Counties), the Middle Rio Grande (Sandoval, Bernalillo, and Valencia Counties), and Socorro–Sierra. To these adjoining constituencies, the real possibility of defaulting on Rio Grande Compact obligations presents a common dilemma, for compact debits and credits apply equally to all three, and a water-budget deficit incurred in one segment affects each of the others. The annual average basin-wide shortfall is currently estimated at 40,000 acre-feet in surface flow; an additional 71,000 acre-feet in aquifer depletions is poised to impact the Rio Grande within the current planning horizon (see paper by Hathaway and MacClune in this volume).

A closer look at regional water plans and budgets is revealing. Although region-specific budgets indicate that each region is operating under a negative water balance (regional consumptive use exceeds regional inflow), the ever-increasing basin-wide deficit is largely a result of unfettered growth and ground water withdrawals in the Middle Rio Grande planning region. Being on the downstream end, the Socorro–Sierra region inherits the Middle Rio Grande valley's debt in the form of reduced inflow, and the region's outflow reflects the projected basin-wide budget deficit of 40,000 acre-feet. To offset their deficit, the Middle Rio Grande planning region proposes to rely, in part, on water purloined from neighboring regions, particularly Socorro and Sierra Counties. The Middle Rio Grande water plan seeks to increase regional supply within the next 50 years by purchasing water rights and ultimately drying 12,500 acres, or approximately one-half of the irrigated cropland remaining in Socorro County, and by transferring salvaged water from 17,500 acres of restored bosque.

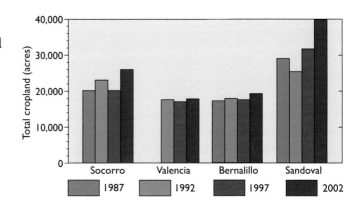

Total cropland in the Middle Rio Grande valley, 1987–2002. Census data indicate that total cropland has remained steady in Valencia and Bernalillo Counties and increased in Socorro and Sandoval Counties, even while water rights are transferred from irrigated lands for other uses. Data is from U.S. Department of Agriculture's 2002 Census of Agriculture, conducted by the National Agricultural Statistics Service.

Implementation of the Middle Rio Grande plan would be both problematic, in a hydrologic sense, and destructive to local interests and commerce in the Socorro–Sierra region, the Middle Rio Grande valley, and the state as a whole. Based on U.S. Department of Agriculture statistics for 2002, the market value of agricultural products associated with 12,500 acres of irrigated cropland in Socorro County is estimated at $17.2 million. It goes without saying that the loss of half of the irrigable farmland remaining in the county would precipitate economic and environmental doom for the region, particularly if the connected water moves with it. Loss of irrigated agriculture of the magnitude proposed would reverberate throughout local retail, commercial, and ranching businesses in the entire Middle Rio Grande valley. The sale and transfer of an agricultural water right currently precipitates either the abandonment of irrigated land or continued irrigation through a water bank. When abandonment occurs on the floodplain, the land eventually reverts to water-consuming wild vegetation, while unregulated water banking leads to a doubling of actual water use. Both courses of action fail to achieve the intended hydrologic response of reducing the budget deficit, and could well wreak profound environmental, economic, and cultural damage. This alarming vision prompted a public welfare statement in the Socorro–Sierra regional water plan encouraging retention of agricultural water rights in the region, and caused the Middle Rio Grande plan to be met with apprehension and outrage by water planning representatives, decision makers, and residents in Socorro and Sierra Counties. The strategy where one jurisdiction tries to achieve water solvency at the expense of a neighbor is generally unpalatable and strictly prohibited in the Interstate Stream Commission's regional water planning guidance. In the final analysis, what does the state of New Mexico do when the Middle Rio Grande region has garnered the entire valley's water yet continues to demand more?

PATHWAYS TO CONFLICT RESOLUTION IN THE MIDDLE RIO GRANDE VALLEY

Substantial conflicts exist between the Middle Rio Grande valley regional water plans, yet the plans themselves and planning forums hold potential solutions. There is no single solution for resolving water and land management issues in the Middle Rio Grande valley, but the following list of remedies could provide benefits:

Public Welfare Statements—State statute requires consideration of public welfare in water rights administration. A number of regions, including those in the Middle Rio Grande valley, have developed public welfare statements, either within regional water plans, comprehensive plans, or county ordinances. Regional public welfare statements were envisioned by the 1986 investigative team and should be considered by the state engineer where water rights transfers may prove detrimental to a region and/or fail to promote hydrologic balance within the basin.

Administrative Solutions—A number of administrative solutions are available to the state engineer to improve water efficiency in the Middle Rio Grande valley: implement active water management in the Middle Rio Grande Conservancy District, declare critical management areas, meter on-farm or ditch diversions, meter domestic and other wells, and require compliance with maximum water conservation standards for all users—urban, rural, and agricultural.

Eliminate Water Management Conflicts within Middle Rio Grande Conservancy District—Two competing water authorities, the state engineer and the Middle Rio Grande Conservancy District, both claim control over management of water within the Middle Rio Grande valley. This competitive approach needs to be resolved so that both water delivery to priority farmers and the state's compact obligations are met.

Agricultural Efficiency and Innovative Cropland Management—Protection of agricultural land for future food production requires that new methods of managing agricultural land to keep a maximum area productive and economically competitive while minimizing water application be developed and applied. For example:

- Spread a water right over a larger area, fallow larger portions of irrigated plots in rotation, and transfer a portion of the right, keeping agricultural lands whole and productive.

- Continue to improve efficiency of water conveyance and on-farm application.

- Increase crop yields and incorporate more high-cash crops in rotation.

- Prevent abandonment of irrigated lands and reversion to invasive water-using vegetation by requiring a portion of any transferable right to remain attached to the land.

Manage Water Markets—Aggressive, open water-transfer markets that currently operate in New Mexico do not adequately balance the statutory mandates of private property right, conservation, and public welfare. There is understandable support for the notion that open markets should be allowed to operate aggressively to facilitate water transfers from agricultural to urban use as a means to accommodate growth and achieve hydrologic balance. However, third party impacts, including adverse effects on rural communities and the environment, should be taken into account, and alternatives that avoid such impacts should be favored. This goal can be advanced through development of criteria that guide water transfers and consider region-of-origin protection and third-party interests.

Conflict Resolution and Collaborative Planning—Pursue resolution of regional and user conflicts through collaborative problem-solving forums. Examples include:

- The Upstream-Downstream Project works toward resolution of regional water conflicts in the Middle Rio Grande valley. The project was conceived and implemented by New Mexico Water Dialogue in 2006 with funds from the McCune Foundation and technical assistance from the Interstate Stream Commission and the Utton Transboundary Resources Center at the University of New Mexico. The project combines small work sessions for regional and local decision makers, support from a gallery of water specialists, and focused problem solving by a technical steering committee to resolve water-use conflicts in the Middle Rio Grande valley and identify fair and balanced solutions.

- The State Water Plan/Regional Water Plan Ad-Hoc Committee, appointed by the Interstate Stream Commission in summer 2003, develops recommendations for resolving differences between the State Water Plan and regional water plans and collaborates on strategies for plan implementation. Three topics addressed have been watershed management, implementing infrastructure needs, and water rights transfer policy.

IMPLEMENTATION OF STATE AND REGIONAL WATER PLANS

By addressing conflicts at a local, grass-roots level through regional water planning and collaborative problem solving, the interregional conflicts predicted by the 1986 study team can be mitigated, resolved, or altogether avoided. Regional planning provides a local perspective on public welfare and ensures that local needs are met, now and in the future. Without local input, any solutions or strategies implemented will be met with antagonism rather than acceptance. The 2003 State Water Plan provides a very clear policy statement promoting the role of regions in forging a water policy that accommodates and protects the diversity of New Mexico's communities and regions, and mandates that the state shall defer to regional guidance. What is lacking now is implementation. Local governments must step up to the plate and take the lead to ensure that regional plans are updated and implemented so that local water security is maintained. The state, through its various agencies, must continue to support and adequately fund update and implementation of both the state and regional plans. We can not afford to stop the process now.

Suggested Reading

State Appropriation of Unappropriated Groundwater: A Strategy for Insuring New Mexico a Water Future. WRRI Report No. 200, New Mexico Water Resources Research Institute and University of New Mexico Law School, 1986.

Water Needs and Strategies for a Sustainable Future. Western Governor's Association, 2006.

Water—Things to Do Now, and Do Better

Frank Titus

I pose a seemingly simple question here: What should we be doing to manage water affairs in the Rio Grande valley (and throughout New Mexico, for that matter) that we are not doing? I ask this penetrating question with some trepidation, because it is enormously complex. The potential number of thoughtful responses is huge, but I offer here a very short list of activities I suggest be elevated immediately to very high priority within the Office of the State Engineer and by the New Mexico State Legislature. If this means delaying other programs, so be it. I list below three specific proposals. Raising the emphasis on these activities is intended to advance our water welfare now, but more importantly to prepare a favorable environment for other more specific changes and improvements, some large, some small, that must subsequently be considered. I follow these proposals with a list of specific actions that would modernize our approach to water management in New Mexico.

1) Bring water rights records up to date in the WATERS database. No task in the Office of the State Engineer (OSE) is more important than compiling and maintaining a set of accurate records as the basis for water rights within the state. Years back the office began to copy its legal records into a computer-based file that would be available on the Internet. This electronic database is called the Water Administration Technical Engineering Resource System, or WATERS. Though current water rights transactions are said to be appearing on WATERS at a measured pace, the comprehensive compilation is far from complete and far from current.

Within this proposed state-wide task, the records for the Rio Grande and its ground water basins must top the priority list. Along this, our greatest river, transfers from surface water to ground water, movement of rights upstream, water-banking claims, and other legal changes are occurring at an accelerating pace. Water rights marketing, solicitations, speculation, and consulting activities (technical, legal, and otherwise) are ever expanding. The validity, legality, and orderliness of all of these depend on accurate and available public records. That is what WATERS was intended to be. Without these records, many water-transfer activities begin to look like a set of shell games. The lack of availability to the public of basic water rights data is unacceptable.

What seems to have happened is that the effort to build WATERS became large enough that the agency has not been able to staff it while carrying on its day-to-day functions. I received a comment many months ago from a staff member of OSE that until adjudication is undertaken on the Middle Rio Grande, the high level of effort needed to build WATERS cannot be sustained. To me this argument is backward: Adjudication cannot be accomplished without the records being in order and available. Such orderliness is a necessary precondition for preparing the myriad individual cases that will comprise the ground work, the negotiations, and the litigation of water rights adjudication.

Water rights trading in the marketplace should not be planned, accomplished, or recorded without accurate, verifiable historical backgrounds for every trade. Without WATERS, neither citizens nor professionals can check their own or their clients' water rights records. An up-to-date WATERS database would also help specialists in the OSE do their jobs. The complexities created by inaccurate or unavailable records extend well beyond those already cited. For instance:

- Many state engineer permits to pump ground water in the past were based on promises to retire surface water rights when pumping began to diminish river flows. Such "dedications" are of questionable legality; they constitute large, poorly defined water rights commitments.

- Water transfer records that affect the Middle Rio Grande Conservancy District, municipalities, counties, and other political entities are not listed in any repository other than WATERS, hence they cannot be publicly checked, audited, or used.

- Double dipping—the practice wherein transferors of surface rights continue to use water after it is transferred—is increasingly common under current administration, and it cannot be controlled or even reliably recognized without complete records.

- OSE evaluates and rules on current water rights transactions one at a time; without WATERS it has no apparent way to access a cumulative set of records for an entire basin.

2) Assure that Middle Rio Grande Conservancy District water deliveries meet priority and other dictates of state water law. The Middle Rio Grande Conservancy District (MRGCD) was created under state law in the early 1920s to reclaim water-logged farmland, provide flood protection in the middle valley, and consolidate the irrigation delivery system in the river reach from Cochiti Dam to Bosque del Apache National Wildlife Refuge. The district is generally bounded on the valley sides by the highline irrigation canals. The district obtained substantial financial assistance in its early years from federal agencies, principally the U.S. Bureau of Reclamation. MRGCD operates with considerably more autonomy than most governmental agencies and claims full authority over water management activities in the Middle Rio Grande valley.

MRGCD is the giant among water agencies in the Middle Rio Grande valley. It presides over water delivery to farming operations that result in about 22 percent of all water depletions in this reach of the valley. The conservancy district's basis for water rights claims is a 1931 OSE permit to change the points of diversion for some 132,000 acres-worth of surface irrigation water from the Rio Grande, including 8,847 acres of Pueblo lands with reserved rights, 80,785 acres of perfected agricultural water rights, and 42,482 acres of non-appurtenant junior rights claimed by the district as a result of salvage through its drainage system. Historically, the maximum non-Indian acreage under irrigation at one time on the Middle Rio Grande floodplain may have been more than 60,000 acres. This marked difference between the amount of non-Indian land ever irrigated and the district's claim of more than 123,000 acres of water rights should trigger careful legal analysis of the discrepancy. This area was traditionally irrigated by 70 or more historic acequias that were subsumed by the water delivery system of the MRGCD. Now, nearly 90 years after its formation, the district's claims to water have never been legally defined nor subjected to the normal constraints of beneficial use. The Office of the State Engineer has repeatedly demanded that MRGCD support its claims by submitting formal documentation for Proof of Beneficial Use. To date the district has not complied, though it may now be nearing completion of such a document.

Another issue that water rights owners within the district should demand be explored legally is whether pre-1907 rights appurtenant to multiple irrigated farms can be accumulated and claimed by MRGCD under some unique form of ownership. Such a claim should raise several issues, including at least (a) whether such landowners have thereby lost their priority positions for water delivery; (b) whether the district can lease or sell water outside of its boundaries when the delivery to any of these farms is short of the vested water right; and (c) whether these conditions are such that no pre-1907 right holder can, as an individual, sell his right outside of the district.

Under its broad responsibilities and power, the district has grown into a mighty but little understood, sometimes impenetrable, agency. It assesses all landowners within its boundaries, not just farmers, and it publicly reports very little of its operational procedures. It appears that only a very small minority of its taxable constituents attempt to understand its operations or decision processes.

In recent years, with the growth of municipal and related demands for water, the sale and transfer of individually owned pre-1907 water rights to non-irrigation purposes has increased. There are clear signs of two common effects from such water rights sales. The first is that MRGCD record keeping, especially records of the "move-from" lands, may not even exist; if it does exist, it is not transparent. The second is that after water rights sales, lands often continue to be supplied with water, through the drilling of domestic wells for subsequently built homes, or by the district, which for a price continues to supply irrigation water from a hypothetical bank of "junior" permitted rights. Such "double dipping" is patently wrong, irrespective of any arguments over legality, and hydrologically unsound in an over-appropriated basin.

The conservancy district has shown little interest in determining how much acreage and which farms have pre-1907 water rights. Furthermore, it has stated a preference for a "parity" (or shared-shortage) philosophy of water management rather than the state's statutory concept of priority of ownership. For MRGCD, non-priority management is advantageous, and certainly simpler than priority management. From the perspective of irrigators with older water rights, however, that simplicity will come at high cost: Their farms could receive little or no water during shortages, whereas, under a priority system, they would be at the head of the line.

It is apparent today that MRGCD's long-term aims are no longer geared solely to serving the farming community. Rather, they now suggest interest in the increased power of being the regional water provider, based on the ability to control delivery to a very large block of water rights. This inevitably will dilute the power of the farming community. In fact, it already is doing so; the district's operational decisions seem dominated by moves to position itself as the major water supplier to municipalities. The water rights that are its capital are all based originally on individual ownership. By claiming that such water rights have somehow reverted to the MRGCD, an immense block of "capital" is created to meet the increasing demands of municipalities. If the district can develop a water-banking system that operates to its advantage, any water rights it claims need never be lost or sold. Rather, they can be leased to municipal governments, thus assuring perpetual dominance over regional water resources.

One should expect that MRGCD, being a water agency formed under state law, would support the Interstate Stream Commission (ISC) in its annual need to send a specified volume of water downstream. Compliance with the Rio Grande Compact is by far the most aggravating and potentially costly problem that New Mexico faces in its water affairs. However, MRGCD's operations and goals probably further threaten the state's basic ability to continue meeting its compact commitments on this river. The state engineer is the best (and likely the only) authority that can solve this problem—by assuring that the district plays by the same water rules as everyone else, and that individual water rights are not assumed summarily to have become the property of the regional water delivery agency.

It must be recognized that the problems of expansive water rights claims by MRGCD are ultimately unsolvable without adjudication. We must move with dispatch toward this legal resolution. Nevertheless, while awaiting adjudication, the district cannot be permitted to continually expand its wet-water use. Once expanded, trying to force contraction will be legally more difficult and doubly painful. Remember, this argument is being played out in an environment already conclusively shown to be short of water for compact delivery.

3) Create formal accounting, reporting, and operating rules for water banks. Formal, secure, well-understood procedures for water banking do not exist in New Mexico. Sale, with permanent transfer of rights, is currently the only structured way to augment water use in one part of a water basin with rights from another part. Such rigidity severely limits whatever beneficial role the marketplace might provide were water availability allowed to adjust to temporal and spacial variations in demand and supply in a given basin. What currently exists is an unorganized mess of "rules" invented individually by entrepreneurs, and private and public would-be water banks. There is little in the present "system" to instill confidence that accurate, auditable record keeping is part of today's water banking, or that protections exist for lessor and lessee, or even that transactions are backed by valid water rights.

I propose formalizing water-banking procedures. Defining such procedures requires creation of two new types of rules: rules to establish operating and reporting procedures for water banks themselves, whether the water banks be commercial entities or government agencies, and basin-specific rules to establish what arrangements are and are not permitted within each individual water basin.

Water banks should be legally recognized and the rules under which they operate standardized by the state. The rules should guide record keeping, auditing, transparency, legal accountability, and procedural standardization. Water banking should have a degree of reliability and security reminiscent of money banking. Its operations should be simple and open. The legislature is the appropriate place for established rules governing water banks. These rules do not have to be complicated, but they do need to be explicit.

Basin-specific rules should have characteristics that reflect both a standard of performance throughout the state and the individual and unique needs of each basin. The staffs at OSE and ISC should cooperatively produce an interim set of rules, then invite detailed input from the public within each basin, including input from the formal public groups that produced relevant regional water plans. The OSE should be the responsible state agency, but if the various rules are well constructed, it should not have to involve itself in individual lease transactions.

Creation of detailed rules for water banks and the state's several water basins will require a significant effort. Interim rules could be expeditiously devised, preferably by small panels of experts, and the rules then tested during an interim period of one or two years. Here are a few water-banking concepts that might be considered in constructing a preliminary set of rules:

- Set limits on the distance upriver or downriver a lease could be transferred.

- Provide some form of area-of-origin protection that addresses third-party impacts.

- To be leasable, a water right should be on record within the OSE system and should be in the WATERS database (to use a pre-1907 right, for instance, it must have been declared).

- A water right should only be leased for up to two years before reverting to its original land base for two or more years (the intent is to disallow permanent leasing).

- Some percentage of the leased water should be tithed to support a specific ecosystem activity or benefit within the basin or area of origin.

- Accurate records must be rigidly required, record formats should be defined, and all leases should be published and transparent.

- The beneficial use for the leased water should be specified in the lease.

- Assure, through a formal tracking system, that water banking doesn't facilitate "double dipping."

Until formal rules are in place, the present systemless arena will continue to invite manipulation and will provide little protection for the rights of participants. This is especially true when the absence of rules is combined with failure of the OSE to provide access to fundamental water rights records. It is easy to anticipate in these circumstances that those most likely to be injured will be the small players, such as individuals and family farmers.

The immediate and equally pressing reason we should be moving to reestablish and simplify the internal order and consistency in water management, however, is the inevitability that change will be forced on us, if and when we cannot meet our Rio Grande Compact commitments. Today we are still free to discuss and devise our own fixes for the inefficiencies and inequities already recognized in the system. But if failure to meet the compact is the driving force, our flexibility goes down, and the ultimate costs go up.

THINGS ARE NOT WORKING WELL NOW AND SHOULD BE CHANGED

An impressive community of citizens has for more than a decade been expressing serious concerns about New Mexico's water future. This very knowledgeable community includes technical specialists on environmental and water affairs and many non-specialists who have learned a lot about environmental and community welfare. The voices of this community, especially those concerned with environmental sustainability, have been raised again and again in forums like the New Mexico First town halls, conferences of the Water Resources Research Institute (WRRI) and Middle Rio Grande Water Assembly, meetings of New Mexico Water Dialogue and the Public Interest Research Group (PIRG), and groups that wrote the regional water plans. Here, in no particular order, are some of the topics that have been discussed repeatedly in these forums and that need to be fixed:

- The State Water Plan must be made into an actual plan to explicitly control our water destiny, and it must be made implementable and enforceable.

- Comprehensive water budgets should be constructed by region. Regional negotiations are impossible without budgets that show instream flows, riparian use, aquifer storage, human uses, etc.

- Water rights priorities should apply to both surface and ground water, hence priority must be equally enforceable on both.

- Establish instream flow requirements. Rivers must explicitly be allowed to have water, and rivers' rights should fit formally into state water rights systems.

- Eminent domain over water rights should be limited. The public should discuss this, and public opinion should have major influence on final decisions.

- Try for a formal, state/tribal agreement on Indian water rights. This will help avoid individual tribal lawsuits, which likely would result in tribal inequities.

- Measure all diversions; require measurement and reporting of all water diversions of both ground and surface water.

- Change state statutes to lengthen the tenure of the state engineer and minimize political turnover. The long learning curve makes rapid turnover inefficient, and even a tough, effective, visionary administrator cannot survive pressure politics.

- Conservation that reduces depletions should be required of all water-use sectors; water saved should go to mitigating the basin deficit.

- Work toward a healthy bosque of native plant species; that means working toward removing exotics and optimizing the mix and density of native species.

- Rejuvenate the valley environment below Bosque del Apache National Wildlife Refuge. Put the river down onto its floodplain and replace exotic phreatophytes. If farming would use less water than wild phreatophytes, find a way to transition to farms, and a way to water them.

- Restrict the extent of Elephant Butte Reservoir so that it does not extend above the Narrows, in order to reduce evaporative loss. Negotiate for replacement storage upstream (e.g., at Abiquiu); offer to share saved water with the Elephant Butte Irrigation District and Texas.

- Establish a research farm on the floor of Elephant Butte Upper Basin to study low-water-use crops, evapotranspiration suppression, and related water-saving technologies. If low-water-use farming is practical, devise cheap leases and water grants or other innovations. New Mexico State University has received substantial federal funding for decades as our Land Grant College and should lead this effort.

- Encourage and support basic and applied research in the varied fields that intersect in the general realm of hydrogeology; to continue improving system management, we require ever-improving levels of technical understanding and ever-increasing data.

- Strengthen levees to a standard level of protection. This is a no-brainer wherever levees are essential for protecting human populations. Here's an imaginative yet pragmatic idea championed by a growing number of ecosystem thinkers: In conjunction with the levee system, create bleed-off areas outside the levees into which floodwaters could be diverted to help take the crest off of high flows, while simultaneously replenishing the aquifer, nurturing fish habitat, and supporting a healthy mosaic of native bosque. The aim would be to re-create some of the beneficial effects of natural system dynamics rather than sticking to old, unimaginative methods of floodwater control. Seek ways to make this permissible under the Rio Grande Compact.

Now, let's get moving. Every thinking New Mexican knows we have water management problems that we don't address. I doubt, however, that most citizens realize the several astonishing discrepancies between reality and management practice that our water leaders have been handed from the past but continue themselves to condone. The agonizing loss of our argument before the U.S. Supreme Court over the Pecos River Compact nearly twenty years ago has required stressfully negotiated adjustments in the Pecos River valley, is costing the state a lot of money, and should point inescapably to the much more stressful and immensely more costly problems we ultimately will face on the Rio Grande. The state engineer's Active Water Resource Management program provides an essential first step toward management modernization. The recommendations in this paper offer another step. Let's hope that the Office of the State Engineer and the state legislature cooperatively elect to use these as a springboard toward hydrologic reality in water management, and toward greater justice for our citizens.

The Unintended Consequences of Water Conservation

Zohrab Samani and Rhonda Skaggs, *New Mexico State University*

Although the phrase *water conservation* means different things to different people, it generally implies an act or policy that will result in additional water for other uses. Conservation of water is widely accepted as good, whereas wasting (or not conserving) water is bad. However, how conservation outcomes are assessed depends upon unit of analysis or point of view, and there is often a discrepancy between the physical reality of the hydrologic system and both public and agency perceptions of water issues. Many people are incorrectly convinced that certain activities (such as increasing agricultural irrigation efficiency) will inevitably result in additional water for other uses. Public policies have been implemented and billions of dollars in public and private investments spent in the name of conserving water in irrigated agriculture. Unfortunately, many of these investments have not made additional water available to new users. In some cases they may result in less water for other users in the basin.

Water conservation is a cultural and political icon that is considered by many to be beyond reproach. In today's highly charged water resource debates, skepticism about water conservation is tantamount to an assault on religious sanctities. Quite often, water conservation *intentions* carry more weight than water conservation *evidence* in policy debates, funding opportunities, and newspaper headlines. This paper addresses the discrepancy between intentions and evidence, as well as the unintended consequences of water conservation, particularly as related to irrigated agriculture in New Mexico.

WATER DEPLETION AND IRRIGATION

Evapotranspiration from the watershed's surface is the true depletion or loss of water from a hydrologic basin. The principle is based on the Theory of the Conservation of Mass; water diverted (i.e., removed from its natural course or physical location through a canal, pipe, or other conduit) and applied in irrigation in excess of evapotranspiration is not lost, because much of it flows back into the basin from which it was withdrawn. This water eventually becomes available to other users at other times in other locations, although a fraction of diverted water in a basin may be unavailable to other users because of incidental losses such as evaporation from open water surfaces and moist soil, non-beneficial evapotranspiration from riparian vegetation, and contamination; because the water returns to the basin too late or too far away to be of practical use; or because the water flows into an irretrievable sink (such as the ocean) or an area beyond reach (such as another state or country).

Irrigated agriculture accounted for 76 percent of total water withdrawals in New Mexico in 2000. It is commonly assumed that reducing water depletion through increased irrigation efficiency will always result in extra water, and agriculture is under pressure to change. However, hydrologic systems are not zero-sum entities where one user's diversion is always another user's loss. The hydrologic reality is that one user's water "inefficiency" often serves as the source of another user's water supply. Examples illustrating this situation are presented below.

DRIP IRRIGATION

For several years farmers have faced a steady barrage of recommendations to use sophisticated irrigation technology (rather than traditional surface irrigation methods) and thus increase on-farm irrigation efficiency. Sprinkler irrigation was an early recommended technology; now drip irrigation is commonly recommended. Drip irrigation allows for precise application of water into plants' root zones, with very little deep percolation loss. There is generally a linear relationship between evapotranspiration and yield over a wide range of crops and water applications. Consequently, irrigation technologies that apply water at optimal times and locations in plant root zones increase crop consumptive use of water and crop yield even as irrigation efficiency increases. For example, because subsurface drip irrigation of alfalfa does not have to be suspended during harvest, the consumptive use of drip-irrigated alfalfa is higher than surface-irrigated alfalfa, where the crop usually experiences significant water stress when harvesting machinery is in the fields.

Alfalfa is grown throughout New Mexico's irrigated areas. The evapotranspiration requirement for an acre of alfalfa is typically three acre-feet of water. This level of consumptive use in the desert Southwest is an example of deficit irrigation, where a crop is irrigated with less water than what would allow the crop to reach its potential yield with full irrigation. Assuming an on-farm irrigation efficiency of 75 percent, the farmer would need to apply four acre-feet per acre of water. Thus, three acre-feet per acre are consumed by the plant, and one acre-foot per acre returns to ground water through deep percolation. This level of evapotranspiration will result in approximately five tons of alfalfa per acre. If the farmer adopts drip irrigation, consumptive use can easily increase to five acre-feet per acre (or more), with potential yields of eight tons per acre (or more). An on-farm irrigation efficiency of 75 percent is actually low by the standards of commercial farms in southern New Mexico, where these efficiencies have been found to be as high as 93 percent (pecans and alfalfa) and 95 percent (cotton) as a result of deficit irrigation practices.

Water "waste" through deep percolation or runoff will be reduced through drip technology, but more water will be consumed by the plant. The individual farmer who uses the technology will have increased yield and income per unit of land. From the farmer's perspective, the new water-conserving technology has had positive effects. However, basin-level consumptive use increased. This does not mean that drip irrigation will always result in increased depletion in every irrigated region. For example, farmers in New Mexico's Las Uvas Valley pump water from a deep aquifer to produce alfalfa with an irrigation efficiency of about 40 percent due to the area's sandy soils. Water lost to deep percolation ends up in a saline clay formation and is not currently recoverable. In this case, drip irrigation would result in saving applied water even though the depletion impact is about the same. Generally, the link between increased irrigation efficiency and reduced return flow is most applicable to shallow, stream-connected aquifers.

IRRIGATION SCHEDULING

Irrigation scheduling involves applying water to growing plants in accordance with their consumptive use needs. Frequency and duration of scheduled irrigations are based on environmental conditions, plant growth stage, and predicted evapotranspiration. Successful scheduling requires knowledge of plant water needs, and an irrigation system that is flexible enough to respond to changing needs throughout the growing season.

Proper irrigation scheduling can significantly increase yields and crop quality. For example, during the nut-filling period in southern New Mexico pecan production (late August to early September), a delay in irrigation can result in large yield reductions. Irrigation water applied to pecans and many other crops does not have the same yield and quality

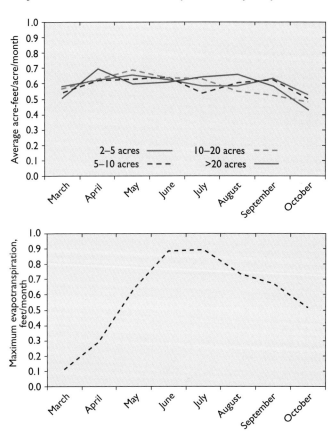

Application of irrigation water to pecans in southern New Mexico by farm size (top panel), and maximum monthly evapotranspiration, in feet per month (bottom panel).

impacts throughout the growing season. With correct irrigation scheduling, total yield and total water consumption by the crop can both increase. If a farmer's goal is to produce a higher level of economic return from a unit of consumptive use, then optimally scheduling irrigations to match crop evapotranspiration is a very desirable practice—but one that may result in increased total evapotranspiration for the basin.

There are approximately 20,000 acres of pecans in southern New Mexico. A recent study of irrigation practices on 340 pecan farms in the region showed that farmers (over a wide range of orchard sizes) were over-applying water during periods of low consumptive demand and under-applying during the critical period of high consumptive demand. This mistiming of water applications results in yield reductions and low water use efficiency (i.e., yield return per unit of water consumptively used). The figure on the previous page indicates that the average application of water is fairly consistent with average crop water needs over the entire growing season, with over-application balanced by under-application. However, irrigation is a means to an end, and pecan yields are compromised by the irrigation patterns shown.

Like efficiency-enhancing drip technology, optimal irrigation scheduling can result in increased consumptive use and subsequent reductions in water available to downstream and future users because of the net reduction in water supplies. The surface water shown as over-applied in the spring months in this figure returns to the hydrologic system, and is no longer available to the trees to which it was applied. The water, which eventually reenters the aquifer or the river, (1) contributes to surface water supplies later in the growing season, miles (and possibly a state or nation) away from where it was originally applied, (2) is likely degraded in quality as a result of salt leaching, and (3) contributes to ground water recharge. Widespread adoption of accurate irrigation scheduling could increase on-farm water use efficiency, yields, and pecan growers' incomes; however, it would also reduce the basin's downstream flows (whatever quality they might be), reduce ground water recharge, and contribute to salt accumulation.

CANAL LINING

Canal lining is often considered to be a "magic bullet" for reducing water "losses." However, canal lining can result in negative water conservation and is unlikely to produce more water for new users. Good examples are provided by the acequia systems of northern New Mexico and the return flow system of the Middle Rio Grande Conservancy District. These systems are essentially man-made Rio Grande tributaries, as they contribute to return flows in the main river channel. In such cases, lining a canal will make farm diversion more efficient but will result in extra depletion if the diverted water is transformed into higher yields. The consequence of this can be reduced in-stream flow, lower water quality, reduced return flows for downstream water users, and overall increased net depletion in the river basin.

The belief that canal lining will conserve water is international in scope. Millions of dollars are spent annually in the United States and other countries to line canals for the purpose of increasing water supplies for urban, industrial, and agricultural users. Canal lining projects are likely to increase gross economic returns for existing farmers or the irrigation district in the affected area. Thus, canal lining may be valuable as a means to increase farm household and district revenues, but it may have no capacity to increase water available for other users. Furthermore, canal lining can negatively impact natural vegetation and wildlife because of seepage reduction.

WHAT IS TRUE WATER CONSERVATION?

Water conservation is basically sound and appropriate water management, but how that translates into specific definitions and actions depends primarily on individual objectives. For a commercial farmer, achieving higher yields through efficient water application is sound water management. For a downstream farmer, water flowing through a drain system as a result of an upstream farmer's over-application is sound water management because it created his water supply. For an advocate of the environment, greater in-stream flows are sound water management, even though those in-stream flows may be the result of very "sloppy" upstream water management. For a growing community with ever-greater demands for recreational opportunities, using water to create a high-quality golf course is sound water management. Dozens more specific examples of sound water management could be given here. Regardless of the number of examples, they all imply that there is and will be little widespread agreement as to what our water conservation objectives should be.

In a water deficit environment such as New Mexico, the technological magic bullets often proposed for agricultural water conservation may have no positive net effect and may in fact increase total basin-level depletions. This fact is rarely advertised by technology vendors and contractors, or by those who advocate the use of public funds to support these technology investments. Adoption of the technologies discussed above is likely to increase depletions and result in less water for other users in a basin. New Mexico's compliance with downstream interstate compact delivery requirements have been and will be compromised by the adoption of such technologies.

If New Mexico intends to get serious about agricultural water conservation in the future, then one of the first steps that should be taken is accurate accounting of basin-wide water use. Accurate accounting would likely release water for new users. Lack of large-scale accounting and administration may be cheap for water managers, but it contributes to an informal water resource management system that relies on ignorance and inadequately defined property rights for its perpetuation. Unfortunately, more data, information, knowledge, and understanding of the water resource are not considered good things by many members of the water resource community. Ignorance resulting from the lack of rigorous water measurement and accounting preserves and protects the status quo, and encourages opportunistic scofflaw behavior by many water users.

Increasing water resource accountability would be a dramatic change in the status quo and would require large increases in data and information about water use. Increased accountability would be a major culture change, and quite different from the popular technological Band-Aid approach (e.g., drip irrigation, irrigation scheduling, or canal lining) to agricultural water "conservation." Of course, agricultural consumptive use in New Mexico can be significantly reduced, and potentially huge volumes of water can be made available to other users, if agriculture itself is reduced in size and land is converted to non-agricultural uses. That change would be dramatic as well and appears to be well underway in parts of the state. However, in New Mexico's current water deficit environment, eliminating water consumption by one user (i.e., agriculture) does not necessarily mean water will be released automatically to other users. Natural vegetation is also a deficit water user, and additional water may simply be consumed by natural vegetation before it becomes available to other potential users.

Suggested Reading

Irrigation Practices vs. Farm Size: Data from the Elephant Butte Irrigation District by R. Skaggs and Z. Samani. Agricultural Experiment Station and Cooperative Extension Service Water Task Force Report #4, New Mexico State University, 2005.

Water Use by Categories in New Mexico Counties and River Basins, and Irrigated Acreage in 2000 by B. C. Wilson, A. A. Lucero, J. T. Romero, and P. J. Romero. New Mexico Office of the State Engineer Technical Report 51, 2003.

Balancing the Budget: Options for the Middle Rio Grande's Future

Deborah L. Hathaway, *S. S. Papadopulos & Associates*

A water budget analysis for the Middle Rio Grande region[1] including both surface water and ground water indicates that, on average, and assuming a wide range of historic climatic conditions, the sum of water used plus New Mexico's Rio Grande Compact delivery obligation exceeds water inflow. This analysis indicates that under present development conditions and without mitigating actions, the region will experience an average annual surface water shortfall of approximately 40,000 acre-feet per year. At present, ground water pumping, which provides the water supply of most cities, imparts an additional annual deficit of 71,000 acre-feet to aquifers; this is the amount of ground water pumping that has not yet impacted the river. Through recent favorable circumstances and management actions, New Mexico maintains compliance with the Rio Grande Compact. Nevertheless, many stakeholders in the Middle Rio Grande understand that the existing pattern of water supply and water use is not sustainable. Meeting existing demand and maintaining compact compliance will become more difficult as the lagged impacts of ground water pumping on the river continue to grow. Further challenging the water planning process are increased water uses associated with projected population increases and maintaining endangered species flows and habitat.

Recent water budget modeling underscores what has been assumed by water management for decades: The basin is "fully appropriated." In fact, the water budget studies suggest that on average, the basin is over-appropriated. New water uses impacting stream flows can only be supported by the cessation of existing uses such that the overall consumptive use of stream flow does not increase, or new water sources must be developed.

WATER SHORTAGE: AGAIN!

Over a period of many centuries, water users have periodically faced the dilemma of water shortage in the Middle Rio Grande region. A casual reader of historic and archaeological accounts can observe the following historic responses to water shortage:

- ***Management and community-based shortage sharing***—Local communities develop shortage-sharing programs, recognizing water as a community resource, implementing management, and reducing usage as necessary for the common good

- ***Enhanced conservation***—Shortage motivates conservation measures that stretch the available supply

- ***Supply augmentation***—New supplies are developed to support economic development

- ***Adversarial confrontation***—When all else fails

These responses to water shortage are age-old. Water management and shortage sharing founded on community property values are a cornerstone of Rio Grande water administration and continue to be practiced by acequia communities. Although many trace acequia customs to practices of medieval Spain and Moorish influence, one may also find examples of this type of management in Pre-Columbian cultures throughout the Americas, including the pueblos of the Rio Grande. There is logic to this precedent, as its widespread use attests. Similarly, enhanced conservation has been practiced for centuries. Check dams on intermittent waterways were built to hold back water for small plots; lands were terraced to optimize the use of available water; small canals were lined with rocks. Many such examples can be found in both Pre-Colombian and Hispanic acequia cultures, in addition to more contemporary conservation measures such as laser-leveling of irrigated fields. These practices conserved the water resource for use by water-based communities and were effective until either climate variations or population pressures stretched the systems too far. Often in such cases supplies were augmented. Along the Rio Grande, Hispanic colonists augmented supplies by constructing stream diversions that weren't necessary for smaller populations. Later, limits

---

[1] The term *Middle Rio Grande region* is used in this article to refer to the area generally between Otowi and Elephant Butte, and is not to be confused with the Middle Rio Grande Planning Region, which occupies a sub-region within the Albuquerque Basin of the Middle Rio Grande region.

on water availability through diversion alone spawned ambitious water storage projects: from Elephant Butte Reservoir in the early 1900s to Heron Reservoir and the San Juan–Chama Project in the 1960s. Remaining limitations motivated large-scale ground water pumping that continues today. For centuries conservation and management of water supplies for the common good have delayed or minimized impacts of shortage, and cycles of shortage have been answered with the development of "new" water sources. Peppered between have been instances of conflict, including the occasional violent confrontation. And, in the past century, priority administration has been applied in some Western states to allocate supply that falls short of demand.

These mechanisms for handling water shortage remain available today. The challenge for regional water planners and decision makers is to decide how to apply the first three mechanisms (management, conservation, or augmentation) to balance the water budget and how to avoid various renditions of the fourth (adversarial confrontation), perhaps in the guise of a court battle or imposition of federal watermaster.

SOLVING THE PROBLEM? THE NEXT FORTY YEARS

State and regional water planning entities recognize that challenges are imminent in the Middle Rio Grande region. We have forestalled the day of reckoning by careful storage of extra water in wetter years, effective flood routing, improved conservation efforts, and through expansion of water supply from ground water pumping. However, these measures are insufficient to avoid water budget deficits at the present level of development, and, clearly, will not be adequate under conditions of increased growth.

To identify methods for balancing the water budget, the regional planning entities largely situated in the Middle Rio Grande basin, including the Jemez y Sangre Planning Region, the Middle Rio Grande Planning Region, and the Socorro–Sierra Water Planning Region, have recently developed regional water plans. The plans include:

- Improved conservation, including urban and agricultural elements

- Reduction of water use from open water and riparian vegetation

- Transfer of water from agricultural to urban uses

- Acquisition of new water supplies, i.e., desalinated water from distant basins or cloud seeding

However, in significant respects, the plans are inconsistent. Considering regional perspectives on agricultural lands:

- The Jemez y Sangre Planning Region would augment their water supply through the retirement of irrigated acreage (amounts unspecified), including lands above Otowi and lands within the Middle Rio Grande Conservancy District below Otowi;

- The Middle Rio Grande Planning Region would augment their water supply through the retirement of 7,500 acres of irrigated acreage in the Socorro–Sierra Water Planning Region and 11,000 irrigated non-pueblo acres within their own region;

- The Socorro–Sierra Water Planning Region, on the other hand, identifies maintenance of the existing agricultural economy and retention of water rights within the region as key regional goals.

Both the Jemez y Sangre Planning Region and the Middle Rio Grande Planning Region look beyond their boundaries for agricultural lands to retire; however, none of the three planning regions welcome the retirement of agricultural lands within their boundaries to serve the urban needs of others.

Considering riparian vegetation:

- The Middle Rio Grande Planning Region would augment their water supply through the reduction of riparian vegetative water use within the Socorro–Sierra Water Planning Region by 17,500 acre-feet per year (they also propose reducing riparian vegetative water use within their own region by a similar amount);

- The Socorro–Sierra Water Planning Region would augment their water supply through reduction of riparian vegetative water use within their region in an amount within the range of 4,000 to 20,000 acre-feet per year.

The elimination of evapotranspiration from riparian vegetated lands is difficult and costly especially in areas where the depth to water is shallow. Successful projects must replace non-native vegetation with lower water using plants, and they must avoid re-colonization and soil evaporation from low-lying valley lands. Both the Middle Rio Grande Planning Region and the Socorro–Sierra Water Planning Region target an amount of land where these projects might be feasible; however, the combined feasibility is questionable. Taking both plans together, restoration of 54,000 acres of riparian vegetation between Cochiti and Elephant Butte is proposed.

Beyond the inconsistency among regional plans for agricultural land retirement, and the questionable feasibility of riparian restoration goals, are challenges of other plan alternatives:

- *Importation of 22,500 acre-feet per year of desalinated water by the Middle Rio Grande Planning Region.* The importation of desalinated water from the Tularosa and Estancia basins to augment supply is a concept not yet realized in the Middle Rio Grande basin. This alternative will bear significant energy costs along with environmental and legal issues.

- *Reduction of Elephant Butte Reservoir evaporation.* The Socorro–Sierra Water Planning Region proposes reducing water demand by 12,000 acre-feet per year through reduction of riparian vegetation in exposed reservoir bottom land. This alternative presents significant engineering and financial challenges.

Setting aside questions of feasibility, if one assumes that the alternatives identified by the regional water plans are implemented within the next forty years, what is the outcome? A water budget analysis has been applied to this question, making only minimal changes to the proposed alternatives to avoid patently inconsistent elements among regions. This analysis, described in the *Middle Rio Grande Water Supply Study, Phase 3*, indicates that under conditions of full implementation, in 2040 (with projected population increases) the surface water deficit is reduced from approximately 40,000 acre-feet per year to 7,000 acre-feet per year, and the ground water deficit is reduced from approximately 70,000 acre-feet per year to 40,000 acre-feet per year. Complete implementation still finds the Middle Rio Grande region in debt, albeit closer to a balancing point. However, in 2040 ground water depletions are again increasing, even with implementation of ambitious conservation and augmentation actions, and with the Albuquerque Drinking Water Project in place. The "solution" to today's shortage provides some relief but sets into motion actions that again place the Middle Rio Grande region onto an unsustainable course.

Most of the analyses described above have been conducted without significant pueblo participation. Pueblo water uses are unquantified and unadjudicated, yet are generally considered senior to non-pueblo uses. The potential for further development of pueblo water resources casts additional uncertainty on the disposition of available water supplies in the Middle Rio Grande region.

Further clouding projections of future supply are questions regarding climate change. Climate modeling predicts reduced water availability in the Southwest, even if precipitation rates remain constant, because of increased evapotranspiration rates due to increased temperature.

SCENARIOS AND ACTION PLANS FOR THE FUTURE

The Concrete Valley

Various names have been proposed for this scenario, for example, "Phoenix" or the "Los Angeles River." We fail to balance the budget, and large-scale ground water mining occurs. The Rio Grande becomes disconnected from the aquifer; river losses are high, and flows are difficult to maintain. Waterways for agricultural delivery must be concrete lined, perhaps piped. Water table conditions can not support a bosque, and only artificial silvery minnow habitat can be maitained. Large areas of the valley are paved, saving water previously used by riparian or agricultural vegetation, and we still manage to make our compact deliveries. (Under current state engineer administration, this is an unlikely, extreme scenario; nonetheless, it could occur by default with lax or ineffective administration.)

The Watermaster's Plan

We try to manage but somehow fail. A series of years pass with unfavorable inflow for meeting our compact obligation. An interstate court battle ensues, and the river is managed by a federal watermaster. Water management goals are defined by decree; incorporation of sub-regional or community-based management options is unwieldy and unlikely. (No action plan is necessary to achieve this scenario.)

The Preferred Scenario

This one is more difficult to describe. However, most stakeholders and planners have some vision of this in mind. They want a future Middle Rio Grande that looks like New Mexico. They acknowledge a desire/need for some growth, though they want the valley to "remain green." They want preservation of environmental and cultural values. The existing state and regional water plans have initiated the process of defining a set of actions and compromises necessary to achieve the preferred scenario, but they are in their infancy. Immediate and sustained implementation of a coordinated action plan is needed for this scenario.

Key elements of the action plan will draw selectively from historic precedent. A few comments on the applicability of these precedents to balance the water budget in today's shortage cycle are noted:

1. *Management and community-based shortage sharing*—Shortage sharing is appropriate, and necessary, at some "community-based" level. Shortage sharing of limited water rights for a given use in a given area is reasonable, for example, within a municipality or within an irrigation district. The difficult exercise is defining the unit suitable for shortage sharing. Planning regions must ask if it is reasonable to shift the burden of their growth to another region, as such an expectation will likely defeat efforts to create coordinated and successful water plans.

Though some might argue that New Mexico's appropriation doctrine and established code of water law and regulation are antithetical to modern resource management, on the other hand, this water allocation system provides an excellent framework for handling water budget challenges. Appropriation doctrine provides use-limited rights to water with protection for established uses, and, as codified more recently, requires consideration of conservation and public welfare. These features are essential to the protection of existing resources and values in New Mexico. Further, New Mexico offers mechanisms for community-based governance where possible. The statutorial incorporation of acequia management customs, and more recently, the crafting of provisions for local management under Active Water Resource Management, for example, as is being drafted for the Lower Rio Grande basin, offer many advantages for community-based water management under shortage conditions. Further, Active Water Resource Management provides a mechanism for controlling water use when regional supply is insufficient. New Mexico benefits from a robust statutory and regulatory framework but struggles with inadequate funding to comprehensively apply the management mechanisms. Rapid adjudication of Middle Rio Grande water rights and strengthened enforcement of permitted conditions of approval (i.e., ensuring that retired lands stay dry) are essential for future management in the basin.

2. *Enhanced conservation*—Conservation involves using less water for a given use and is an important element of all water plans. However, from a basin-wide water supply standpoint, it is only a *reduction in consumptive use* that stretches the water supply. There are many conservation measures that reduce diversion needs but have no impact on consumptive use. In such cases, diversions and return flows are reduced, but the actual loss of water to the system, overall, remains the same. These measures, though useful from a water operations or environmental standpoint, don't address the problem of basin-wide water shortage. For example, the use of drip irrigation reduces the amount of water needed for diversion, but also reduces the amount of water returned via subsurface drainage to drains and back to the river. Aside from possible reductions of surface evaporation, there is no net water savings. Similarly, water-saving fixtures may save water in terms of inflow, but wastewater returns are similarly reduced. Implementation of such conservation measures won't yield the region-wide savings needed to balance the water budget.

There are conservation proponents who believe that the entity implementing conservation should be entitled to use the "saved" water as an incentive to conserve. This viewpoint is inconsistent with the appropriation doctrine, which provides for a reasonable quantity of water to satisfy a specific use. If the saved water only derives from reduced diversion and a reduced return flow, there is no "new" water to support expanded consumptive use. However, when conservation results in reduced consumptive use, for example, clearing water-consuming brush from ditch banks, the saved water reasonably belongs to the public. In the case of an over-appropriated basin such as the Middle Rio Grande, this type of savings in consumptive use is what is needed to balance the budget. Regional water planners must focus on identifying and implementing conservation measures that reduce consumptive use, and avoid expensive conservation measures that have no net impact on the water budget.

3. *Supply augmentation*—Supply augmentation becomes more difficult with each cycle of shortage. For centuries, water users in the Middle Rio Grande have looked for and found means of supply augmentation to solve shortage (diverting water from streams, pumping ground water, importing water from other river basins). These solutions increasingly come with unintended consequences, the most dramatic being the lagged impacts of ground water pumping that creates a debt for future generations. Prudent manage-

ment would suggest that new sources, if they exist, should be identified, tested, and developed before assuming that they will be available to satisfy future growth.

4. Compromise—Departing from historic precedents, confrontation is replaced with compromise. For balancing the water budget, little "low hanging fruit" remains. Largely, tough choices involving trade-offs remain. Planning regions have initiated the process of identifying trade-offs but will need to follow through with implementation. Adjustments to accommodate changed conditions, including additional pueblo uses or climate-based supply reduction, will be required. Actions that can be accomplished within planning regions, without assumption that the resources of neighboring regions are available, are likely to be most successful. The State Water Plan will need to track and reconcile the regional goals and actions. Administration of water rights will require more capital and labor, as without careful monitoring and enforcement, solutions will be circumvented.

In summary, balancing of the water budget in the next forty years, particularly given projected growth rates and climate change impacts, is an ambitious proposal. Success in this endeavor will require focused design and implementation plans, inter-regional coordination, state leadership, political support, and capital outlay, beginning now.

Navigating the River of Our Future—The Rio Poco-Grande

William deBuys

This article first appeared in Natural Resources Journal, v. 41 no. 2, Spring 2001. It is reprinted here with permission.

All elements of the southwestern landscape—deserts, prairies, woodlands, and forests—are much changed from their aboriginal condition, and they continue to change. Unfortunately, society's ability to recognize and adjust to those alterations invariably lags the changes themselves. We are slow to define the dimensions of change, slower to agree that it demands adaptation, and slowest of all in implementing needed adjustments, which are nearly always complex and difficult, requiring new political consensus and institutional change.

For a cautionary example of how badly our society has handled such challenges in the past, we need look no further than the story of the western range and the debacle of overgrazing that transformed it. As early as the 1870s, John Wesley Powell and others began advocating a system of leases to avert a continued tragedy of the commons on the rangelands of the public domain. But the on-going crisis produced little but argument for more than half a century. Between 1899 and 1925, eighteen bills to regulate grazing on the public domain were proposed in the US Senate; in roughly the same period the House entertained twenty-five such bills. All failed, largely because of fear that regulated leases—instead of wide-open winner-take-all competition—would enable big operators to squeeze out homesteaders. The effect of this stand-off was to assure the continued deterioration of a vital resource long after the resource was acknowledged to be in danger. Lack of action to abate this tragedy persisted until enactment in 1934 of the Taylor Grazing Act, which survived its crawl through Congress thanks to clouds of Dust Bowl dirt raining their proof of national ineptitude on the nation's capital. Even so, the problems of the western range were hardly cured.

The present overstocked, fuel-heavy condition of western forests is another sobering example of society's faltering ability to adapt to the environmental changes it engenders. In this case, decades of fire suppression have produced conditions that favor stand-changing fires of an intensity unprecedented in the natural history of the forests most affected. An increasing frequency of catastrophic fire has galvanized popular desire to address the problem, enough so to cause the ship of public policy to begin a long, slow turn. The pilot house of that ship, however, has proved to be a crowded and argumentative place, and agreement on the new course the ship should follow remains elusive.

The fate of southwestern rivers, like that of the region's forests and rangelands, is being shaped by society's response to the most fundamental problem affecting its relationship to the environment. This is the competition between the survival needs of complex ecosystems, on the one hand, and the task of providing natural resources for human use and essential services such as flood protection and waste disposal, on the other. We are linked in this task to our forebears. Previous generations worked with great resolution and energy to develop the resources of New Mexico and the Southwest. Our generation now faces the obligation to deal with the consequences of that development.

This, in a sense, is a leading theme of the history we are making today. Since 1492 most of the environmental history of North America has involved, literally and figuratively, the breaking of new ground. But as we encounter limits of supply (as with western water) and adverse consequences of past use (as with forest health), the history of the future will increasingly involve contending with the consequences of what was broken. For the sake of prosperity, if not survival, we and our neighbors throughout the world have entered an age of obligatory adjustment and repair. This is one of the fundamental tasks of our time. It's not what our fathers and mothers, grandfathers and grandmothers undertook, but it is our mission, and history will judge us on how well we accomplish it.

These themes of alteration, competition, and both the difficulty and necessity of repair dominate the history of the Middle Rio Grande. If we were to look at a map of the alluvial plain of the river as it existed in, say, 1900, and if each vegetation type within that corridor—agricultural field, cottonwood forest, marsh, oxbow lake, flood scour, etc.—were differently colored, the result would be an intricate and vivid mosaic, a close twining of many different habitats, sprawling several miles wide across the valley floor. Moreover, this image of diversity would be dynamic not just in space, as captured by the map, but also in time. In 1900 the riparian corridor was an environment in rapid motion. Year by year and season by season, the mosaic changed as the river flooded, abandoned old channels, adopted new ones, and repeatedly altered ecological conditions in one location after another. From a human point of view, such a system was messy, chaotic, and frequently dangerous. It was also very inefficient in terms of providing steady, predictable, capturable outputs.

If we were next to look at a similar map for the year 2000, we would see a much simpler image. Nearly all of the riparian habitats are now restricted to a narrow corridor between levees, while agriculture and, in many areas, urban and suburban development dominate everything else. Significantly, the contemporary situation is simpler in time as well as space. The frequent shift from one vegetation type to another typical of conditions in 1900 has now become rare with the flow regime of the river tightly managed and its available floodway constrained by levees, the dynamism of the overall system has slowed to a comparative halt.

ORIGINS OF THE MODERN RIO GRANDE

To understand the transformation of the river over the past century, we need to understand the kinds of challenges faced by earlier generations of New Mexicans. By the 1870s more than 120,000 acres were under cultivation along the Middle Rio Grande, but that number soon began to decline because of upstream developments. Over-grazing, cut-and-run logging, extensive fires, the extension of roads and trails (which contributed to arroyo formation), and other factors vastly aggravated erosion throughout the watershed. The net effect was to increase greatly the river's sediment load.

During the 1880s and 90s, meanwhile, Mormon settlers brought much of the San Luis Valley of Colorado under cultivation by opening scores of new irrigation diversions on the uppermost reaches of the Rio Grande and its tributaries. These diversions had the effect of reducing downstream flows, so that not only was the river forced to carry more sediment, but it had less water with which to flush the sediment through the system. Settlement of the San Luis Valley prompted severe water shortages hundreds of miles downstream, and in the late 1880s the Republic of Mexico complained bitterly about the loss of flows at El Paso Del Norte—the area of today's Ciudad Juarez and El Paso. Agriculture there had declined by at least 50 percent, and many families were forced to abandon the area altogether.

The people of the Middle Rio Grande suffered from these changes in another way. The over-burdened and under-watered river was aggrading—the level of its channel was slowly rising due to the deposition of sediment. This loss of channel capacity made the river more prone to flooding and raised local water tables, waterlogging adjacent fields and making them more vulnerable to salinization. Because of these changes agriculture steadily declined along the Middle Rio Grande from the previously mentioned high of 120,000 acres to only 40,000 acres in the 1920s.

Society and its institutions responded. The problems of the Rio Grande, mirrored in watersheds throughout the West, helped spur development and acceptance of a new conservation ethic, which historian Sam P. Hays has aptly called the "Gospel of Efficiency." Conservationists like Gifford Pinchot, Elwood Mead, and W. J. McGee, working with Theodore Roosevelt and others, addressed themselves to the problem of harnessing and harvesting natural resources, especially rivers for irrigation and forests for timber, in order to meet the long term demands of a growing population and its increasingly industrial economy. Pursuit of this task resulted in the creation of new institutions organized to manage key lands and waters. The idea was to protect resources both from hasty and wasteful exploitation by profiteers and from piecemeal, uncoordinated development by interests to small or too inexpert to optimize their usefulness.

But the apostles of efficiency did not stop there. They sought not just to cure society of wastefulness but to purge nature of it as well, and by improving nature, to provide at the highest level more of everything society wanted, water and pasture for agriculture, timber for industry and ultimately recreational opportunities for increasingly urban population.

The apostles of efficiency viewed nature as a large machine, like a factory. The same scientific principles that rendered the factory floor more productive would also make the machine of nature more efficient. The first thing to do was to eliminate waste and superfluous movement, which was accomplished by removing unneeded parts. Among the parts to be removed were floods in rivers, freshwater flowing to the sea, fire in forests, bark beetles and budworms, predators, prairie dogs and other varmints, even porcupines. Granted that a lot of other cultural imperatives entwined with the impulse to simplify, but the impulse remains the common thread. Today we are dealing with the results of those removals in virtually every ecosystem we attempt to manage. Having removed floods and abundant water from the "machinery" of our rivers, we now struggle to keep the Middle Rio Grande and a few similar survivors alive, while others, long dead, we treat as ditches.

INSTITUTIONS FOR THE NEW RIVER

Regardless of how we may value those removals today, it is instructive to look at what it took for them to occur. What was required for society to make so radical a change in the way it tended its land and water? First came recognition that ecological and social conditions had changed. Next came development of a social and political consensus that action was necessary. And finally it became necessary to form new institutions to execute the necessary action. In the case of the rivers, the needed institutions would provide flood control, drainage and irrigation, and minimize the instability of watersheds through forest and range management.

The history of most of those the institutions is fairly well known. In 1902 the Reclamation Service was created, and it grew into the Bureau of Reclamation. Construction of Elephant Butte Dam had begun in 1903 under private sponsorship but was soon suspended because of Mexican protests. The Reclamation Service, eager to show what it could do, took over the project and completed it in 1916. The U.S. Forest Service came into being in 1905 and in the years thereafter asserted management control over much of the Rio Grande's forested watershed.

Not all of the new institutions were federal. In 1925 New Mexicans created an institution to implement the gospel of efficiency in the valley of the Middle Rio Grande. By action of the state legislature, they formed the Middle Rio Grande Conservancy District (MRGCD), and they gave it the power to condemn acequias and to levy taxes (although the district scrupulously avoids calling them taxes).

The Middle Rio Grande Conservancy District replaced some seventy acequia headgates and ditches with four major water diversions feeding an area-wide system of high-line canals, distribution channels, and drainage ditches. Soon the new system began to achieve the economies of scale and higher efficiency in agriculture that its backers had been hoping for. There was, however, a counter current of conflict between the larger interests that profited from the new regime and the older, smaller operations that struggled to meet the higher level of capital investment required by the new system. The smaller interests protested, but they did not prevail.

Ultimately, a second set of conflicts having to do with interregional river apportionment were settled, and a permanent Rio Grande Compact became the law of the river in 1938. Two new dams supported implementation of the agreement: El Vado in 1935 and Caballo in 1938.

Then came the river's last great flood. Flows of 25,000 cubic feet per second inundated towns along the river, including Española and downtown Albuquerque. The floodwaters broke through the MRGCD's levees and destroyed much of the infrastructure that the district had built. Damage to property in Albuquerque was terrific, not least because the aggradation of the Rio Grande had brought it to a level higher than the city's downtown area. The disaster of 1941 led to congressional concern and attention, but action to correct the situation had to wait until after the conclusion of World War II.

In 1948 Congress approved the Middle Rio Grande Project and authorized the Bureau of Reclamation to dredge and channelize the river, to reconstruct levees, and to confine the river from meandering with gabions and jetty jacks and other means armoring the channel. To accomplish this work, the bureau entered into an intricate relationship with the MRGCD, a partnership that has continued to evolve, notwithstanding that the two partners have not always agreed on its terms and conditions.

Through the post-war years the bureau and the Army Corps of Engineers, in one of the unhealthiest bureaucratic competitions of all time, sought to outdo each other in building dams throughout the West. The Rio Grande did not escape their attention. Platoro was completed in 1951, Jemez Canyon in 1954, Abiquiu in 1963, Galisteo in 1970, and Cochiti in 1975.

Additionally, New Mexico finally got its share of both pork and water from the Colorado River Compact. The pay-off took the form of the San Juan–Chama project, which has authority to divert up to 94,000 acre feet of the Navajo River into the Rio Grande watershed by means of a tunnel through the continental divide. This project required its own dam and reservoir, and Heron Lake came into being in 1971.

Much of this mightily expensive new plumbing was built in the name of agriculture, but interestingly irrigated fields in the middle valley today occupy about 54,000 acres which is only about 14,000 acres more than it did in the 1920s. Use of valley lands for settlement, commerce and industry has of course produced much greater transformations and unquestionably the biggest economic impact of river control and engineering has been an enormous creation of wealth in terms of real estate value.

ECOLOGICAL CHALLENGES

Where previously the Rio Grande meandered over a flood plain that in some areas was miles wide, today we have a tightly constrained river and a small floodable area between the river levees. The transformed river is far more stable, more reliable, and more efficient than its predecessor in providing the resources and services that society identified as its highest priorities early in the last century. But the transformation has also produced an unwanted decline in ecological diversity and health, a good deal of which is attributable to the fact that the river has not flooded meaningfully in decades. This is a profound change, for river floods represent the single most powerful force in structuring the riverine and riparian environments.

The idea of a "structuring force" warrants expansion. These days we are well aware of our obligation to take care of certain elements within specific ecosystems. In the forest, it may be the spotted owl or the Jemez Mountain salamander. But taking care of individual species is hardly easy. In recent decades ecologists have learned that attempts to maximize individual variables in a complex, multi-variant system (which every ecosystem is) tend to cause the system to falter or crash. It doesn't seem to matter what the variable is. It might be board feet of lumber or animal unit months of grazing. It might be deer or codfish. The lesson seems to be that if the system managed singlemindedly for the production of one output, the overall system tends to decline, often precipitately.

On the other hand, it is impossible to attempt to manage every element of an ecosystem, for there are far too many of them. In most cases it is impossible to identify them all. Contrary, perhaps, to most people's expectation, this healthy realization does not leave us without alternatives. It leads us instead to acknowledge that, instead of trying to manage individual variables, we have to focus on trying to release the *keystone processes* that structure and shape a system. Where rivers are concerned, flooding is a keystone process. Or more accurately, the keystone process is the river's natural hydrograph—its flow regime, embracing the full pattern of high and low flows and their variability through time. In ponderosa pine forest, low-intensity, frequent fires are among the keystone processes that structure the forest system. If an ecosystem can be managed to permit its keystone processes to function in a naturalistic pattern and at a naturalistic intensity, then the individual variables, be they commodities or endangered species, will tend to take care of themselves—and persist at a sustainable level. This is a main truth that seems to be emerging these days from our experience as a scientific culture in dealing with complex ecosystems.

To return to the Rio Grande, the formerly dominant cottonwood gallery forest of the river's riparian corridor generally requires flooding to reproduce. In the absence of flooding, cottonwoods lose their competitive advantage to other plants including the Russian olive, Siberian elm, salt cedar, and other species. We are accustomed to calling these plants "invaders" as though they were launching some kind of assault. But what they are doing is not invading, they're simply making use of a heavily modified habitat that we've created and that welcomes them by meeting their needs for establishment and reproduction. Among these better adapted plants, Russian olive and Siberian elm dominate the bosque understory in the upper reaches of the Middle River where the river is degrading, while tamarisk, or salt cedar, dominate the riparian zone in the lower reach of the middle river where the Rio Grande is aggrading.

Informed visitors to the Rio Grande Nature Center in central Albuquerque will quickly note that the forest ecosystem beside the river is undergoing rapid change. They will see a vigorous understory of Russian olive lining the riverside drain. Cottonwoods provide the topmost canopy, but nearly all of these tall trees hail from the Class of '41, the last great flood and the last year of extensive cottonwood reestablishment. One looks in vain for young cottonwoods in the understory but finds instead Russian olive, tamarisk and other exotics. It is the Russian olive, not the cottonwood, that is reproducing most successfully, while the many of the older cottonwoods are senescent. Visitors who look carefully at what is on the ground, will feel their concerns grow even more acute. In most areas one finds heavy accumulations of dead wood, for there has been no flood to carry it off nor any standing water to saturate the material and speed its decomposition. One need not be a forester or ecologist to sense that the bosque in the vicinity of the nature center is unnatural-looking and very much in peril. It is ready to ignite from the first Roman candle on the Fourth of July or from a dropped cigarette or lightning strike. The native cottonwood/willow riparian systems of the Southwest are not well adapted to fire, but we have made them extremely vulnerable to destruction by fire.

Unless the management paradigm shifts, the great cottonwoods—and by extension, what we think of as the native bosque of the Rio Grande—will continue to decline and eventually perish. Not that a riparian community will cease to exist. There will always be trees and other plants growing along the river, but the riparian community of the future will be quite different from that which evolved in concert with the river.

The traditional bosque, dominated by native vegetation, touches many a cultural nerve. As a society, we've come to value things that our forebears, who harnessed the river in service to other values, took for granted. Those formerly abundant things that are now scarce or threatened, and hence dear, include open space, woods, access to the river's edge, opportunities for recreation, solitude, contact with nature, and many living traditions both Indian and Hispanic.

The bosque also possesses an aesthetic dimension that deserves our attention, one that those who consider themselves defenders of the river would do well to bear in mind. Those who would have society change its ways should not expect to succeed by making a case based solely on fact. One has to appeal to the heart, too. And so it is important consider the beauty of the bosque—and the way the native system speaks directly to the heart. The cottonwood is an icon of the West. It's arching canopy offers shelter and shade, in a land where both are scarce. Its furrowed bole stands fast against restless skies. Most important, the cottonwood signals water amid dryness. And its fat leaves, the size of a child's hand, applaud the slightest breeze with a sound like rain. In *Great River*, Paul Horgan described the main road along the Rio Grande as, "passing in and out of cottonwood shade, a river grace." That grace, along with the bosque of the Middle Rio Grande, is today a rare and vanishing thing.

Something else that we value today, again because of its increasing scarcity, is bio-diversity, and by extension, ecological health and vigor. The river provides ample evidence that the news is not good from this quarter. Lunkers like the shovelnose sturgeon, the grey redhorse, and the freshwater drum were gone from the river by the end of the last century. Soon afterwards, the American eel also disappeared. It came as a surprise to me to learn that the common eel, which breeds in the Sargasso Sea out past the Caribbean, migrated all the way to the upper Rio Grande. We know this because the Tewa at Santa Clara and San Juan used eel skin in certain of their leggings and ceremonial dress. But eels could not make the trip after Elephant Butte Dam went into place.

THE SILVERY MINNOW

Meanwhile, at least four species of cyprinid fishes, the large family that includes carp and sunfish and most freshwater minnows, were extirpated from the Rio Grande between 1949 and the late 1960s. Probably many interacting factors contributed to the loss of these species, but chief among them were alteration of the river's hydrograph and reduction of streamflow resulting from irrigation withdrawals and reservoir storage. The diminished flows, combined with the drought of the 1950s, effectively dried up large stretches of the river for longer stretches in time and distance than had been the case before. Construction of levees and channel manipulation also simplified the river laterally, eliminating prospects for sloughs and ponds that might have functioned as refugia during times of low flow. Two of the vanished species are believed to be extinct. Two others can still be found in portions of the Rio Pecos. All of them had the bad judgement to depend upon scarce water in limited habitat, a trait shared by the last of the river's endemic cyprinids, the Rio Grande silvery minnow (*Hybognathus amarus*) which clings tenuously to life in the river whose name it bears. Once abundant from northern New Mexico to the Gulf of Mexico, the minnow clings to existence in no more than 5 percent and probably less than 1 percent of its original habitat. Unfortunately it is making its last stand in the stretch of the Middle Rio Grande most vulnerable to drying—an indication that habitat modifications in wetter stretches, including possibly the presence of exotic predators, may be even more inhospitable to the minnow than low water levels.

The silvery minnow was officially listed as an endangered species in 1994. In the following year, the minnow's terrestrial neighbor, the southwestern willow flycatcher, a typically drab and jittery *Empidonax* flycatcher that all but the most expert birders find impossible to distinguish, joined the minnow among the unhappy elect of the endangered list. The flycatcher's plight reflects a decline of the riparian environment in exactly the way that the minnow stands for the decline of the river.

These considerations compel us to view the future of the Middle Rio Grande and it's bosque with grave concern. But the context of the problem is even graver, and its severity cannot be overstated. As bad as the news may seem to be for the hammered ecosystem of the Middle Rio Grande, it describes the best situation existing on any reach of a major river in the entire Southwest. Ecologically, the native communities of the Middle Rio Grande may be on their last legs, but all other comparable systems are either prostrate or defunct. Estimates of the loss of native riparian habitat in the Southwest range from 85 to 98 percent. There is little doubt but that the bosque of the Middle Rio

Grande, among all its kindred ecosystems, is the best remaining example.

A PATH FORWARD

If the people of central New Mexico were contrarian enough to do what their neighbors in Arizona, California, west Texas, and northern Mexico have decidedly not done—that is to say, if New Mexicans elected to maintain the Middle Rio Grande and its bosque as a live river instead of a dead ditch—what would they do?

First, they would have to accept that they can't get the old river or the old bosque back. They can try to maintain a new kind of Rio Grande and a new kind of bosque within more or less present limits of constraint, which are set by the levee system. Although opportunities may exist for moving the levees back in a certain locations (notably in the southernmost reaches of the Middle Rio Grande), surrounding development makes these opportunities few. In protecting this minimalist riparian zone, river managers would not recreate a "natural" system. Indeed, in a land than has supported heavy human use for many centuries, it would be hard to choose an appropriate model for what that natural system might have been. A more reasonable goal should instead be to create a context in which naturalistic processes might continue to operate. This, I submit, is the kind of nature Americans might best hope to encounter in any of the landscapes they tend—forest, desert, riparian, or grassland. Even so, this is tall order.

Friends of the river would also accept that the future of the bosque will include exotic species. Salt cedar may not be native, and programs to control it may reduce its dominance in certain areas, but it will always be with us. So will Russian olive, Siberian elm, tree of heaven, and others. The same applies to the exotic river fishes that have found homes in the system—they are analogs to the waves of human arrivals that have swelled the population of the Southwest in recent generations.

The top priority for environmental river management is to keep the native elements of the river system present and to manage the system to allow operation of the keystone processes that favor those elements. Furthermore, we need to be prepared to accept and even promote patches of disturbance. A healthy riverine system will be a vigorous and dynamic mosaic of early, middle, and late successional gallery forests, plus ponds and lagoons, wet meadows, sandbars, and scour areas. This kind of continuous renewal is often hard for people to accept. When we see something we like—a stand of towering cottonwoods, for instance—we're inclined to say, "hold it right there, don't change a thing." But we ignore at our peril the truth that nothing holds still for long, not our children, ourselves, nor the manifestations of the natural world around us. If we really want to keep the things we value most, we have to learn to roll with the system's inherent dynamic of change. This is especially important when we deal with the exceptional dynamism of a riparian system like that of the Middle Rio Grande.

The key to management of a renewed Rio Grande will be to manage the plumbing of the river, including all its dams, drains and diversions, to mimic as closely as possible the natural hydrograph of the river. This means that flows would spike and fall in the seasonal pattern that characterized the river's behavior before it was dammed. The hydrograph is the keystone process we most need to honor. A central element of such an effort would be to arrange as often as possible for overbank spring floods, still within the levees, to promote regeneration of cottonwoods, to speed decomposition and recycling of nutrients, to carry off or dampen understory fuels, and so forth. It is probably not important for us to enumerate all the things floods do for the system. In fact, we probably cannot catalog them all, anyway. We mainly need to know that the system works much better with them than without them.

Having floods, incidentally, requires having a levee system capable of accommodating and withstanding high flows. There is no small irony here that levees are important, not only to protect us from the river, but to protect the river from our diminishment of it. Within the levees, the river can perhaps be permitted to behave like a river.

These prescriptions for keeping the river and the bosque alive are presented in an unusual study, of which any serious student of the Rio Grande should be aware. This interagency study drew on the resources and expertise of the U.S. Fish and Wildlife Service, the Bureau of Reclamation, and the Army Corps of Engineers, under the direction of professor Cliff Crawford, of the University of New Mexico. It was completed in 1993 and is generally known as the Bosque Biological Management Plan. Essentially, this study identifies the principal management goals that must be achieved if the Middle Rio Grande is to remain a live river with a surviving native bosque. Understanding those goals is a vital first step, but the hardest work still lies ahead.

The most daunting thing in the middle river is not understanding the natural ecology of the river. The most daunting problem is its contending with its political ecology. The fate of this long, thin ribbon is controlled by no less than four counties, nine towns and cities, six pueblos, four federal agencies, five state agencies, the Interstate Stream Commission, and the Middle Rio Grande Conservancy District. The complexity of our contemporary world makes it relatively easy to hobble or stop complicated undertakings. With consensus difficult to achieve and veto power widely shared, it is infinitely harder to resolve complex matters, even when such resolution promises to benefit all affected parties.

True conservation management of the Middle Rio Grande would indeed be complex. The challenge is to get all the interested entities, all of the complex political ecology, working in the same direction. In the end, the choice between a landscape embodying complex or simple nature, between a live river or a dead ditch, will come down to choices about use of water. Maintaining remnant cottonwood and willow bosque via flooding, for example, raises the question of whose water will be used for the flood and where will that water go? Having floods means having a place to put the floodwaters after they run through the system we wish to treat. This may prove to be a significant obstacle to restoring floods to the system.

Rivers like the Rio Grande need "prescribed floods" in the same way many forests and grasslands need prescribed fire. In both cases the prescription is to restore a keystone process. One place where prescribed floods have been used effectively and recently, is on the main stem of the Colorado. Intentional high volume releases from Glen Canyon Dam in 1996 produced a salutary effect on the ecology of the Grand Canyon. Hopefully, the means will be found to continue that kind of practice in the future. But on the Colorado, the managers had the huge capacity of Lake Mead, downstream, to absorb those waters.

The Middle Rio Grande lacks that kind of capacity. Under certain circumstances Elephant Butte Lake may serve to accommodate floodwaters, but sedimentation has greatly reduced its original capacity, and interstate and international agreements make its management far from flexible. Moreover, every drop of water in the river is spoken for. In a sense, so extensive is the accounting of water rights on southwestern rivers that every drop in every stream is owned by someone—or more accurately, a succession of someones—even before it falls from the sky as snow or rain. It used to be that floods were considered an act of God. One might say that the "waste" of water was charged to His account. But after nearly a century of dam and levee building in the spirit of the gospel of efficiency, only the most extraordinary weather conditions today produce volumes of water that exceed the capacity of the system to control, store, and mete out according to plan.

Nowadays most floods must necessarily be acts of man. In a watershed, where every acre-foot of water is allocated and owned, even before it exists, even before the water molecules that comprise it are deposited in the form of rain or snow, the intentionally permitted disappearance of water must theoretically be debited to someone's account. Exceptions do exist, but they are small and rare.

Floods do not occasion the only or even the greatest need for the allocation of water to environmental purposes. Maintaining minimum flows in certain habitats to sustain endangered species can consume much larger amounts of water. In either case, a clear need exists to create what we might call "water entitlements" for individual rivers like the Rio Grande. This is a decidedly post-modern concept, reaching far beyond the ordinary bounds of irony. To argue that rivers are entitled to a share of the water they carry might in other times and places seem unnecessary but not in the contemporary Southwest, where neither the laws nor practices of the past century and a half are sympathetic to any but a utilitarian view of the waterways that make our oasis civilization possible.

Perhaps one day, such thinking will seem as strange to our successors as belief in the divine right of kings now seems to us, but in the meantime, those who would endow our rivers with water must find that water within the existing legal and administrative system. One obstacle to doing so is the prevailing myth that western water allocation is a zero-sum game—that all water is fully and precisely allocated, subject to water rights defined with crystal clarity, and that the systems that use this water run with the precision of a Swiss watch. According to this line of thinking, any reordering of such a system will blow its precision to smithereens.

The reality, however, is quite different. Look closely at any cluster of water rights and uses, and you quickly learn that uncertainty abounds. Who owns exactly what? Who has used how much water, and for how long? Precise answers to such questions turn out to be surprisingly hard to come by. On the Middle Rio Grande, for instance, the water rights of the MRGCD have never been definitively quantified—notwithstanding that the district is now nearing eighty years of age. Besides the existence of many conflicting and competing paper claims to water, a lot of water is used inefficiently. Every system leaks. Most systems operate as much on assumptions as hard data, and only rarely are those assumptions entirely correct. This is not a

Swiss watch. It is more like a sundial on a partly cloudy day.

Precisely because there is flex in the system, opportunities exist to secure water for the Middle Rio Grande without injury to current holders of water rights. Public agencies, like the Bureau of Reclamation, already lease water on the open market from willing sellers like the city of Albuquerque. At some time in the future one or more pueblos along the Rio Grande may choose to disentangle their water rights from those held by the MRGCD and similarly lease water for environmental purposes.

Donations might also be made. A municipality like Albuquerque might voluntarily contribute water to a river entitlement as a means of inspiring greater water conservation among its citizens.

The biggest opportunities, however, lie with agriculture. The long-term trend throughout the West is for agriculture, which uses approximately 80 percent of all water, to become more efficient, and for the water thereby saved to be reallocated to urban and industrial uses. When such reallocations occur, a portion of the redirected water should be reserved for environmental protection. Urban purchases of water can pay for conservation infrastructure—field leveling, drip systems, canal lining, computerized transmission control, etc.—so that agricultural production and economic activity does not diminish. Similarly, federal or state funds might pay for infrastructure that frees up water for endangered species protection. Or funding might directly pay for fallowing or forbearance during drought years in order to provide water for minimum flows.

The mutually beneficial alternatives, while perhaps not profuse, are nonetheless not scarce. The greatest obstacle to progress in maintaining the life of the Middle Rio Grande is the reluctance of vested interests to contemplate them with an open mind.

Inaction, however, is not the only danger. Obedient to the law of economics, the managers and constituents of our hypothetical water entitlement would seek to maximize use and minimize costs of the system and endeavor to buy, donate or require only as much in the way of rights for ecological uses as is considered absolutely needed. How much is that? Unfortunately, such a question can never be fully resolved, for it can only be answered in terms of current knowledge, on which full agreement never exists.

(Indeed, it's reasonable to question the security of all our assumptions about water availability. Most data indicate that during the quarter century from 1970 to 1995 we enjoyed one of the wettest pluvial periods in recorded history. Indeed, if we consult prehistoric measures, we seem to inhabit the wettest period of the last two thousand years. This realization should inspire an ethic of restraint in the society of the Southwest, but it goes largely ignored.)

A GLOBAL CONTEXT

The literature on a wide range of attempts to manage sustainable harvest of natural resources—acre-feet, board feet, animal unit months of grazing—surfaces three fatal problems that lead virtually all of these attempts to failure. The three themes or characteristics include:

- An inevitable push to maximize economic returns

- The use of operational models built on current knowledge, which is never complete and on which there is never full agreement

- The lack of full agreement on the scientific "facts of the situation," which effectively throws decision making into the political and economic sphere and further guarantees overuse

These implacable conditions describe the bleak endgame of water allocation in the American Southwest, and every man, woman, and child who lives within the compass of the Rio Grande is in it.

The question facing New Mexicans is this: Will we place ourselves enough ahead of the curve of thirst and urgency to create a buffer and to tithe a portion of the water with which we are blessed to the system that provides it? Such a tithe need not be made purely from a sense of moral obligation. There can and should be self-interest in such an act. Anything set aside, any flex in the system, anything that is not allocated to current consumption and use, becomes a buffer against uncertainty. It becomes a buffer against the droughts and other surprises the future inevitably brings. Saving something for the river, saving something for the system, creates a reserve from which people will also benefit. Not least, it checks the development of ever higher and less sustainable levels of dependency.

In contemplating their water future, the people of the Middle Rio Grande would do well not to lose sight of two ideas: the first is that the MRGCD, by far the region's greatest consumer of water, is not the problem. The MRGCD is the solution. The MRGCD, together with state and federal agencies, should pursue an aggressive and thorough examination of its operations to determine where and how water might be saved by achieving a higher level of efficiency. Having determined how to save water, the district can then calculate how much the savings will cost and it can begin selling the saved water at a reasonable rate, while in the process upgrading its infrastructure to a more efficient, easier-to-manage state. It can reduce operating costs by selling water for the cost of the capital to save that water.

The second idea this: the silvery minnow, the endangered species that has engendered so much litigation and debate over the use of water in the Middle Rio Grande, is not a curse on the region. It is a blessing. At present, that tiny fish, only a couple of inches long, is the only thing that prevents the Rio Grande through Albuquerque from becoming like the Rio Grande through El Paso. Or the Salt River through Phoenix, or the Santa Cruz through Tucson. I will wager that nearly every person who reads this article has been to one or more of those cities. And I will wager that none of them made an effort to notice where those rivers were. That is because there's nothing much to notice. They are ditches, nothing more. Unless the paradigm of management changes, the Middle Rio Grande may suffer a similar fate.

The silvery minnow may save the region from itself. It is there to remind everyone—city-dwellers, farmers, tribes, and visitors—that we owe more to the places we inhabit than simply the pursuit of the next increment of profit and convenience.

It is worth remembering that all attempts to use or alter the land are attempts to tell a story about how we think the land ought to be. What we find over and over is that these stories we tell are inevitably simpler than the land itself. We cannot escape from geography; we are embedded in it. And the test of our character, as a people embedded in geography, is how well we keep our stories current—how intelligently and effectively we respond when we learn that they are out of date and that they require revision. The question before all of us who depend in one way or another on the Middle Rio Grande is whether we will revise our story about life in this place in a way that keeps a living river in it.

This is the story of many people. It is their *history*, working itself out, day by day and home by home. This class of problem, which involves choosing between an accommodation of complex nature and the relentless pressure of economic compromise, is the problem of the people of the Middle Rio Grande as much as it is the problem of people anywhere in the world. How those of us who are alive today resolve this problem—or fail to resolve it—will afford future historians many insights about the character of our time and place and about the kind of people we have chosen to become.

Suggested Reading

Salt Dreams by W. E. deBuys, University of New Mexico Press, 1999.

Conservation and the Gospel of Efficiency—the Progressive Conservation Movement, 1890-1920 by S. P. Hays, University of Pittsburgh Press, 1999.

Barriers and Bridges to the Renewal of Ecosystems and Institutions by S. Light, L. Gunderson, and C. S. Holling, Columbia University Press, 1995.

Great River: The Rio Grande in North American History by Paul Horgan, Wesleyan University Press, 1991.

List of Contributors

Bland, Douglas M.
Special Projects Manager
New Mexico Bureau of Geology and Mineral Resources
New Mexico Institute of Mining and Technology
76 Encantado Loop
Santa Fe, NM 87508
(505) 466-6696 Fax: (505) 466-3574
dmbland@comcast.net

Doug Bland has worked for the New Mexico Bureau of Geology and Mineral Resources since 2004, primarily as project manager for educational programs and conferences on natural resource topics, including decision-makers field conferences. He served as director of the Mining and Minerals Division of the New Mexico Energy, Minerals and Natural Resources Department from 1998 through 2002, where he was responsible for overseeing environmental protection and permitting of mine sites. He also held various technical and managerial positions in the Mining and Minerals Division between 1989 and 1998. His experience includes twelve years in the mining and petroleum industries. He holds B.S. and M.S. degrees in geology from Virginia Tech and the University of Wyoming.

Bowman, Robert S.
Professor of Hydrology & Department Chair
Department of Earth and Environmental Science
New Mexico Institute of Mining and Technology
801 Leroy Place
Socorro, NM 87801
(505) 835-5992 Fax: (505) 835-6436
bowman@nmt.edu

Robert Bowman teaches and performs research in hydrology and water chemistry in the hydrology program at New Mexico Tech. Much of his career has involved investigations of the fate of contaminants in soil and ground water, including movement of nutrients and pesticides below irrigated fields, leaching of metals from contaminated soils, and transport of solvents and fuels from spill sites. More recently he has turned his attention to surface water/ground water interactions in the Rio Grande valley. He is one of the scientific leaders of a statewide effort, funded by the National Science Foundation, to improve estimates of evapotranspiration in the Rio Grande riparian corridor. Prior to joining New Mexico Tech in 1987, he spent five years as a soil scientist with USDA's Agricultural Research Service in Phoenix. He holds an A.B. degree in chemistry from the University of California, Berkeley, and a Ph.D. in soil chemistry from New Mexico State University.

Cleverly, James
Research Assistant Professor, Hydrogeoecology Group
Department of Biology, University of New Mexico
MSC03 2020
1 University of New Mexico
Albuquerque, NM 87131
(505) 263-9536 Fax: (505) 277-6318
cleverly@sevilleta.unm.edu

James Cleverly is a research assistant professor in the biology program and heads the Hydrogeoecology Group at the University of New Mexico. Dr. Cleverly has led and participated in numerous projects evaluating bosque water use in Nevada, New Mexico, Colorado, Kansas, and Nebraska. His primary research is the study of physiological and ecological relationships between salt cedar and native riparian species, water relations, growth, carbon partitioning, competition, invasion of riparian habitats by salt cedar, evapotranspiration, restoration water salvage, climate, and water loss from the Middle Rio Grande's shallow aquifer to the atmosphere. He recently contributed to the regional water plan prepared for the New Mexico Mid-Region Council of Governments and the New Mexico Water Assembly, providing an evaluation of water salvage due to bosque restoration using state-of-the-art estimates of plant water use at various locations along the Middle Rio Grande. Partnerships that have supported this research include the National Aeronautic and Space Administration, the U.S. Fish and Wildlife Service, the New Mexico Interstate Stream Commission, the U.S. Bureau of Reclamation, and the National Science Foundation EPSCoR program. Dr. Cleverly has a B.S. in biology from the University of Utah and M.S. and Ph.D. degrees in plant physiological ecology (biological sciences) from the University of Nevada, Las Vegas.

Dahm, Clifford N.
University of New Mexico
Department of Biology
Albuquerque, NM 87131
cdahm@sevilleta.unm.edu
(505) 277-2850

Cliff Dahm is a professor at the University of New Mexico. His research in New Mexico has focused on stream and river ecosystem ecology, ground water/surface dynamics, biogeochemistry, geomicrobiology, ecohydrology, and restoration ecology. Cliff grew up in Idaho, where he worked and recreated in the wilderness areas of central Idaho. He presently teaches ecosystem studies, freshwater ecosystems, geomicrobiology, and limnology at the University of New Mexico. He received his Ph.D. in oceanography and aquatic ecology from Oregon State University. He also holds an M.A. in chemical oceanography from Oregon State University and a B.S. in chemistry from Boise State University.

deBuys, William
1413 2nd Street #6
Santa Fe, NM 87505
(505) 984-2871
wdebuys@earthlink.net

William deBuys is a historian and conservationist based in Santa Fe. His most recent books include *Salt Dreams: Land and Water in Low-Down California* (University of New Mexico Press, 1999) and *Seeing Things Whole: the Essential John Wesley Powell* (Shearwater Press, 2001). From 1991 to 1993 he chaired the Rio Grande Bosque Conservation Initiative.

Dello Russo, Gina
Ecologist
U.S. Fish & Wildlife Service
Bosque del Apache NWR
P.O. Box 1246
Socorro, NM 87801
(505) 835-1828 Fax: (505) 835-0314
gina_dellorusso@fws.gov

Gina Dello Russo is the refuge ecologist at the Bosque del Apache National Wildlife Refuge. Currently her duties include planning and implementing riparian restoration projects on the floodplain of the Rio Grande within the refuge (10 river miles) and coordinating with private landowners. Gina is also focal to the Save Our Bosque Task Force efforts on the 45 mile San Acacia reach, as well as collaborative programs on the Middle Rio Grande (160 river miles). She works extensively with other agencies, non-profit organizations, and private citizens to control invasive species and improve habitat diversity, the efficiency of water use by the natural system, habitat for endangered species, and fire protection. She has a diverse background in surface & groundwater hydrology, geology, and biology including 20 years of field experience on the Rio Grande in northern and central New Mexico. She is a graduate of the University of New Mexico where she studied biology with an emphasis on ecology and environmental science.

Ellis, Lisa M.
Department of Biology, University of New Mexico
Natural Connections
604 Wellesley Drive NE
Albuquerque, NM 87106
(505) 268-3944
lmellis@swcp.com

Lisa Ellis is the science coordinator at the Bosque del Apache NWR and one of the principal authors for the Bosque Education Guide and is a co-founder of the Bosque Ecosystem Monitoring Program. She runs a private consulting business doing various monitoring and educational projects, primarily along the Rio Grande. Lisa was the project manager for a long-term study of artificial flooding at the Bosque del Apache NWR. Her dissertation research focused on flooding and fire in the Middle Rio Grande bosque. She holds a B.A. in ecology and evolution from the University of California, Santa Barbara and M.S. and Ph.D. degrees in biology from the University of New Mexico.

Flanigan, Kevin
New Mexico Interstate Stream Commission
Springer Square Building
121 Tijeras NE, Suite 2000
Albuquerque, NM 87102
(505) 764-3865
Kevin.flanigan@state.nm.us

Kevin Flanigan is a senior hydrologist with the Rio Grande Bureau of the New Mexico Interstate Stream Commission. He has approximately 20 years of experience in hydrology and water resources engineering and water rights administration, with the majority of that experience in the Rio Grande basin of New Mexico. He has been with the Interstate Stream Commission for eight years, where his current responsibilities involve water resources management activities on the Middle Rio Grande focused primarily on ensuring compliance by New Mexico with the Rio Grande Compact. This includes review of reservoir and water operations by the U.S. Army Corps of Engineers, the Bureau of Reclamation and the Middle Rio Grande Conservancy District, management of basic hydrologic data collection activities and river maintenance and flood control activities, and review and analysis of various water planning studies and reports and water use plans and projects. He has a B.S. in civil engineering from the University of Michigan and an M.S. in hydrology from the New Mexico Institute of Mining and Technology. He is a registered professional engineer with the state of New Mexico and has been certified as a professional hydrologist by the American Institute of Hydrology.

Follingstad, Mary Helen, AICP
Executive Director
Santa Fe Regional Planning Authority
142 West Palace Avenue
P. O. Box 276
Sante Fe, NM 87504
(505) 995-6508
mfollingstad@co.santa-fe.nm.us

Mary Helen, a native New Mexican, was appointed as the executive director for the Santa Fe City and County Regional Planning Authority in June of 2006. Prior to that time Mary Helen managed the New Mexico Interstate Stream Commission's regional and statewide water planning programs from 1997 to 2006 and was the senior community planner for Santa Fe County from 1983 to 1997. She has been a member of the American Institute of Certified Planners (AICP) since 1989. Mary Helen has completed the Leadership Santa Fe course and is active with New Mexico First. She also volunteers with the Museum of New Mexico Women's Board, is the recent past president of the Tano Road Association, and is a member of the Santa Fe Extraterritorial Zoning Commission. Mary Helen holds a B.A. degree in fine arts from the University of Denver, M.A. degrees in fine arts from the University of Colorado and Saint Johns College in Santa Fe, and an M.S. in community and regional planning from the University of New Mexico.

Hall, G. Emlen
Professor of Law
School of Law, University of New Mexico
1614 1/2 Bayita Lane, NW
Albuquerque, NM 87107
(505) 277-2866
hall@law.unm.edu

G. Emlen Hall is a professor at the School of Law, University of New Mexico, where he teaches water law and edits the Natural Resources Journal. He has written two books on water issues, *Four Leagues of Pecos: A Legal History of the Pecos Grant from 1800 to 1936* (1984) and *High and Dry: The Texas–New Mexico Struggle for the Pecos River* (2002) published by the University of New Mexico, as well as many articles on the state's land and water. Prior to joining the UNM law faculty in 1983, he spent seven years at the Office of the State Engineer. During his time there, he wrote an administrative history of the Pecos River Compact from its inception in 1949 to 1974. This was the beginning of his research for *High and Dry*. When Hall first arrived in New Mexico in 1969, he wrote for and edited the *New Mexico Review*, a monthly investigative journal. He also practiced law in Pecos and served as village planner, attorney, and municipal judge for the Village of Pecos. He has worked for Northern New Mexico Legal Services and the New Mexico Land Grant Demonstration Project. Hall brings a background in water law and public land law to his teaching and writing. Professor Hall holds an A.B. from Princeton University and a J.D. from Harvard University.

Harding, Benjamin L.
Principal Engineer, Hydrosphere Resource Consultants
1002 Walnut, Suite 200
Boulder, CO 80302
(303) 443-7839
blh@hydrosphere.com

Ben Harding is a consulting engineer with Hydrosphere Resource Consultants in Boulder, Colorado. He has been practicing water resources engineering for more than 35 years, and during that time has had a diversity of assignments. Mr. Harding has conducted water availability studies at scales ranging from Boulder Creek to the entire Colorado River Basin. He was one of the principal investigators in the Severe Sustained Drought Study, which examined the water supply consequences of unprecedented drought in the Colorado River Basin. Mr. Harding has served as an expert witness in original jurisdiction interstate compact litigation and large toxic tort litigation. Recently, he has been working on projects to improve short- and long-term water supply forecasting for water providers and to allow planners to quantify uncertainty in estimates of future water supply. He has been a registered engineer in Colorado since 1979. Mr. Harding received his B.S. degree from the University of Colorado.

Harner, Mary J.
Post-doctoral Research Associate
University of Montana
Division of Biological Sciences
Missoula, MT 59812
mary.harner@mso.umt.edu
(406) 243-2393

Mary Harner is a post-doctoral research associate at the University of Montana. She completed her Ph.D. in biology in 2006 at the University of New Mexico, where she was a fellow in the Freshwater Sciences Interdisciplinary Doctoral Program. Her research focused on influences of flooding on riparian ecosystems along the Middle Rio Grande. Mary grew up along the Mississippi River in Alton, Illinois, and has spent much of her life exploring rivers. Her research focuses on riparian ecosystem ecology, with an emphasis on interactions between plants, soil, and shallow ground water. She holds an M.S. in environmental studies from the University of Montana, where she conducted research in association with the Flathead Lake Biological Station on the ecology of cottonwood trees, and a B.S. from Tulane University in ecology, evolution, and organismal biology.

Harris, Steve
Executive Director, Rio Grande Restoration
HCR 69, Box 3-C
Embudo, NM 87531
(505) 751-1269 or (505) 770-2502
Fax: (505) 776-1443
unclergr@laplaza.org

Steve Harris directs Rio Grande Restoration, a non-profit stream flow and watershed advocacy group that he founded in 1994. Steve is also the owner-operator of Far-Flung Adventures, a river outfitting company based on the Rio Grande. His experience as a river guide has enabled him to observe the workings of the Rio Grande first-hand. He has been a member of such private river protection efforts as the Rio Grande Alliance and Forgotten River Advisory Group and serves such public initiatives as the Middle Rio Grande ESA Collaborative Program's Water Acquisition and Management subcommittee, and the New Mexico Strategic River Reserve legislation and regional water planning group for Taos, Santa Fe, and Albuquerque. He resides in a riverside cottage in Pilar, from which he studies, speaks and writes about the history of the river and promotes awareness of the importance of the Rio Grande to people, communities, and ecosystems. Steve received a B.A. in journalism from the University of Oklahoma.

Hathaway, Deborah L.
Vice President, Hydrologist
S. S. Papadopulos & Associates
1877 Broadway, Suite 703
Boulder, CO 80302
(303) 939-8880 Fax: (303) 939-8877
dhathaway@sspa.com

Ms. Hathaway is a principal of S. S. Papadopulos & Associates, Inc., a water resource and environmental consulting firm, and has managed their western office since 1994. She is a hydrologist and water resource engineer with experience in modeling ground water and surface water systems, water rights, water-supply development, contaminant transport, ground water remediation, and regional water planning. She was lead investigator on the Middle Rio Grande Water Supply Study (2000, 2004) conducted for the Army Corps of Engineers and the New Mexico Interstate Stream Commission to support regional water planning in the Middle Rio Grande. Her experience with hydrology and water rights in New Mexico dates back to 1982. Ms. Hathaway worked for New Mexico State Engineer Steve Reynolds from 1982 to 1988. She earned a B.A. in liberal arts at St. John's College in Santa Fe, New Mexico, an M.A. in secondary science education from the University of New Mexico, and an M.S. in civil engineering (hydrology and water resources) from Colorado State University in Fort Collins, Colorado.

Johnson, Peggy S.
Senior Hydrogeologist
New Mexico Bureau of Geology and Mineral Resources
New Mexico Institute of Mining and Technology
801 Leroy Place
Socorro, NM 87801
(505) 835-5819
peggy@gis.nmt.edu

Peggy Johnson is a senior hydrogeologist with the New Mexico Bureau of Geology and Mineral Resources. She has twenty years of consultant and research experience in ground water hydrology and related fields. Her diverse background includes practical research in arid basin hydrogeology, karst hydrology, mountain-front recharge, surface water and ground water resource assessments, isotope hydrology, and water resource management and policy. During her 11-year tenure at the Bureau of Geology, Ms. Johnson has managed or contributed to regional hydrogeologic studies across north central New Mexico, including Placitas, the Espanola Basin, the Albuquerque Basin, and the Taos Valley. She is currently heading a hydrogeologic research project in the southern Sacramento Mountains aimed at improving the hydrologic balance in high-elevation watersheds. Ms. Johnson also serves on numerous water planning and policy committees and commissions, including the Socorro–Sierra Regional Water Planning Committee, the Upstream/Downstream project, the Interstate Stream Commission's Ad Hoc Committee for Regional and State Water Planning, and the New Mexico Water Quality Control Commission. She received her B.S. in geology from Boise State University (Idaho) and her M.S. in hydrology from New Mexico Tech.

Kelly, Susan
Associate Director, Utton Transboundary Resources Center
University of New Mexico School of Law
MSC11-6070
1 University of New Mexico
Albuquerque, New Mexico 87131-0001
(505) 277-0514
skelly@law.unm.edu

Susan Kelly is the associate director of the Utton Transboundary Resources Center at the School of Law, University of New Mexico. The Utton Center is a policy center created to help parties sharing a water resource to manage the resource within a legal framework instead of litigating over it. The center promotes the equitable and sustainable management and use of transboundary resources by providing impartial expertise and scholarship. The center approaches projects from a multi-disciplinary standpoint and also provides educational programs on legal issues concerning natural resources. Kelly is active in a variety of projects, including development of an adjudication water rights ombudsman program for the New Mexico courts, modeling alternative reservoir management scenarios through involvement in Middle Rio Grande Endangered Species Act issues, regional water planning and others. She represents Governor Richardson on several committees working on U.S.–Mexico border water issues. She is an attorney licensed in the state of New Mexico and also a member of the American Institute of Certified Planners. Her law degree is from the University of New Mexico and her undergraduate degree is from Arizona State University. She was formerly the water rights manager for the city of Albuquerque and is a long-time resident of New Mexico.

Klise, Geoff
620 Lafayette Dr. NE
Albuquerque NM 87106
(505) 284-2500
gklise@sandia.gov

Geoff Klise is a contractor at Sandia National Laboratories currently working with a collaborative group from southwest New Mexico to develop a decision support model for the Gila and San Francisco Basins. He has assisted the Utton Center with papers on transboundary water issues and geographic information system support for the U.S.–Mexico border region. Prior to starting his masters program, he worked as an environmental consultant focusing on soil and ground water remediation in Washington and California and worked as a water resources hydrogeologist for the Washington State Department of Ecology reviewing applications to appropriate surface and ground water. He is licensed as both a geologist and hydrogeologist in Washington State. He received a Master of Water Resources from the University of New Mexico and a B.S. in environmental and engineering geology at Western Washington University.

Kreiner, Dick
100 Tavalopa Drive
Las Lunas, NM 87030

Dick Kreiner grew up on a dairy farm in Michigan. After graduating from high school he worked on the family farm for a couple of years before entering the Air Force. When he completed his 4 years with the Air Force, he went to college in Arizona and graduated from the University of Arizona with a Bachelors Degree in civil engineering in 1977. He then got a job with the Corps of Engineers at the Albuquerque District Office. His entire career was focused on water management. He gradually worked his way up to the reservoir control chief and held this position from 1987 to 2003. During his last year and a half before retirement he was the project manager representing the Corps of Engineers in the Middle Rio Grande Endangered Species Collaborative Program. He retired in January 2005. Dick always tried to balance flood control responsibilities with the management needs for human and riparian communities below Corps of Engineers reservoirs.

MacClune, Karen
Project Hydrologist
S. S. Papadopulos & Associates, Inc.
3100 Arapahoe, suite 203
Boulder, CO, 80303
(303) 939-8880
kmacclune@sspa.com

Karen MacClune is the vice president and principal hydrologist at S. S. Papadopulos & Associates in Boulder, Colorado. Karen conducts and directs a wide variety of hydrologic, water supply development, environmental, and ground water remediation projects and has over 12 years of experience designing and conducting basin-scale hydrologic studies. Her experience includes: application of water budgets and hydrologic models to address water planning questions, to evaluate water supply alternatives, and to assess ground water remediation requirements; compilation and organization of data to support conceptual and modeling analyses; quantification and/or estimation of water budget components; and application of past and projected future climatic data to questions of water availability and demand. Karen's current work on projects involving stakeholder communications, ground water modeling, water planning, climate change, and conjunctive surface water/ground water use in the Southwestern US provide her with broad perspectives on community and regional water issues. She has also worked for the Technical Division of the New Mexico Office of the State Engineer in Santa Fe, New Mexico and as the water resource engineering specialist and the water resource specialist in their Water Rights Division. Karen also worked for the Water Resources Division of the U.S. Geological Survey in Santa Fe, New Mexico. She earned her B.S. in mathematics at the Massachusetts Institute of Technology, and her M.S. in geology and Ph.D. in geophysics at the University of Colorado in Boulder.

McCord, Jim, P. E.
Principal Hydrologist and New Mexico Manager
Hydrosphere Resource Consultants
P.O. Box 445
Socorro, NM 87801
(505) 835-2569 Fax: (505) 835-2609
jtm@hydrosphere.com

In his 25 year-professional career, Dr. McCord has worked as a staff engineer for a geotechnical engineering consulting firm, as an assistant professor at Washington State University, as a senior member of the technical staff at Sandia National Labs (SNL), as hydrology group leader with D.B. Stephens and Associates, and (since 1999) with Hydrosphere Resource Consultants, managing their NM operation from their office in Socorro. With Hydrosphere, he has been involved in numerous water resource projects throughout New Mexico and Colorado, and his clients include the NM OSE, NM ISC, Valencia County, the pueblo of Isleta, the city of Boulder, the Colorado Attorney General, and several law firms across the western U.S. He co-authored the textbook *Vadose Zone Processes*, published in 1999. Dr. McCord received a B.S. from Virginia Tech and an M.S. and Ph.D. from New Mexico Institute of Mining and Technology. Jim and his wife Cecilia operate a 28 acre certified organic farm in Polvadera, and he is a founding board member for Rio Grande Agricultural Land Trust, dedicated to preserving open lands and wildlife habitat in central New Mexico.

Mitchell, Matthew
Rio Grande Agricultural Land Trust
Office of the Mayor
One Civic Plaza N.W., 11th floor
Albuquerque, NM 87102
riobirds@zianet.com

Matthew Mitchell is board president of the Rio Grande Agricultural Land Trust, currently serves on the board of the Save Our Bosque Task Force, and is president of the New Mexico Falconer's Association. He has been a raptor rehabilitator with Wildlife Rescue of New Mexico, Inc. for 25 years. Matt and his wife still operate the southwestern jewelry manufacturing business they started while at UNM. He built a home in rural Valencia County and later relocated to rural Socorro County near the Rio Grande and Bosque Del Apache NWR. He originally became involved with restoration of his own river-front property and then with local groups concerned with habitat restoration and limiting of new development in the Rio Grande valley. Matt aspires to protect as much riparian habitat and farmland from loss to development as he can while establishing a mosaic of native riparian habitat along the middle Rio Grande wherever feasible. Matt is a native New Mexican with a B.S. in biology from the University of New Mexico.

Mussetter, Robert A.
Mussetter Engineering, Inc.
1730 South College Ave, Suite 100
or P.O. Box 270785
Fort Collins, CO 80525
Cell: (970) 224-4612 Fax: (970) 472-6062
www.mussei.com
bobm@mussei.com

Bob Mussetter is a river engineering consultant with nearly 30 years of experience on a wide variety of rivers throughout the U.S. and internationally. His expertise includes the integration of hydraulic engineering, sediment transport theory, and fluvial geomorphology to address river engineering issues. He has completed and continues to work on a wide variety of issues throughout the Rio Grande system that involve flood capacity, water delivery, sediment transport, and solutions to associated problems that affect public safety and the health of the ecosystem. Bob is a registered engineer in ten states, including New Mexico. Aside from his consulting activities, he has served on a variety of technical committees dealing with river issues, including a recent assignment as a member of the Adaptive Management Forum Scientific and Technical Panel that was formed by the California Bay-Delta Ecosystem Restoration Program (CALFED) to provide guidance to local scientists, engineers, and managers in finding and implementing appropriate means of restoring the ecosystem in tributaries to the San Joaquin and Sacramento River Basins. Bob has a Ph.D. in hydraulic engineering from Colorado State University.

Parody, Jennifer M.
Middle Rio Grande ESA/Rio Grande Silvery Minnow Coordinator
U.S. Fish and Wildlife Service
2105 Osuna NE
Albuquerque, NM 87113
(505) 248-6657 Cell: (505) 239-8352
jennifer_parody@fws.gov

Jennifer Parody has been with the U.S. Fish and Wildlife Service since 2004, first serving as species lead for the southwestern willow flycatcher, and since 2005 for the Rio Grande silvery minnow. Her primary responsibilities include coordinating Section VII consultations on the Rio Grande, representing the service to the Rio Grande Endangered Species Act Collaborative Program, and coordinating recovery activities for the silvery minnow. Prior to joining the U.S. Fish and Wildlife Service, she worked for the New Mexico State Land Office as their state biologist and in San Francisco with a private consulting firm. Both her master's and doctoral research examined songbirds and large-scale landscape processes. For her dissertation she worked on southwest rivers studying the habitat requirements of Bell's Vireo. Jennifer has a B.S. in resource policy and ethics from Cornell University, an M.S. in conservation biology and ecosystem management from the University of Michigan, and a Ph.D. in biology and evolutionary ecology from the University of New Mexico.

Price, L. Greer
Associate Director/Chief Editor
New Mexico Bureau of Geology and Mineral Resources
New Mexico Institute of Mining and Technology
801 Leroy Place
Socorro, NM 87801
(505) 835-5752
gprice@gis.nmt.edu

Greer Price is associate director at the bureau, where he directs the publications program. His experience includes seven years as a geologist working in the oil patch, ten years with the National Park Service, and four years as managing editor at Grand Canyon Association. He has served on the boards of the Publishers Association of the West and the New Mexico Geological Society Foundation. His focus for many years has been on the interpretation of geology for the general public. His career has involved teaching, writing, and field work throughout North America. He is the author of *An Introduction to Grand Canyon Geology*, and has a B.A. and an M.A. in geology from Washington University in St. Louis.

Robert, Lisa
35 Miguel Road
Los Lunas, NM 87031
(505) 865-1455
elksedge@earthlink.net

Lisa Robert is a native of Albuquerque's South Valley and a "perennial student" of the Rio Grande. She has covered state and local water issues for the past twenty years, publishing an independent newsletter for constituents of the Middle Rio Grande Conservancy District since 1987, and serving as editor of the *New Mexico Water Dialogue* from 1994 to 2001. She is the author of *The Middle Rio Grande Bosque Biological Management Plan Update: The First Decade*, a review of recent history and water policy, and their ecological consequences for the middle basin. She also lovingly farms five acres of floodplain near the historic community of Tomé.

Rosacker-McCord, Cecilia
Executive Director
Rio Grande Agricultural Land Trust
P.O. Box 40043
Albuquerque, NM, 87109
(505) 270-4421
ceciliam@sdc.org

Cecilia owns and operates a certified organic farm in Polvadera, New Mexico, which serves a number of Albuquerque's fine restaurants, grocery stores in Santa Fe and Albuquerque, as well as farmers' markets in Socorro and Albuquerque. She has served on the board of the New Mexico Farmer's Market Association for ten years and has served on the governing council for the New Mexico Food & Agriculture Policy Council for five years. In addition to farming, Cecilia is currently the executive director of Rio Grande Agricultural Land Trust, a non-profit organization dedicated to preserving New Mexico's family farms and ranches, open space, and wildlife habitat for future generations. She holds a B.S. from New Mexico Tech.

Samani, Zohrab
Professor, Civil Engineering
New Mexico State University
Box 30003, MSC 3CE
Las Cruces, NM 88003
(505) 646-2904 Fax: (505) 646-6049
zsamani@nmsu.edu

Zohrab Samani is a professor of civil engineering at New Mexico State University. His teaching and research interests include irrigation, hydrology and hydraulics, applications of remote sensing to water resource management, and alternative energy sources. Dr. Samani works frequently with the Winrock Foundation on irrigation management problems in central Asia.

Schmidt-Petersen, Rolf
Rio Grande Bureau, New Mexico Interstate Stream Commission
Office of the State Engineer
407 Galisteo Street
Bataan Memorial Building
P.O. Box 25102
Santa Fe, NM 87504-4102
(505) 827-6125

Rolf Schmidt-Petersen is the manager of the Rio Grande Basin Bureau of the New Mexico Interstate Stream Commission. His responsibilities on the Rio Grande include investigation, development, conservation, and protection of the Rio Grande water resources and stream system, interstate stream compact administration and compliance, and resolution of interstate and federal water resource issues affecting Rio Grande water resources. Projects conducted by the Rio Grande Bureau include working with Reclamation to construct and maintain the 22-mile long pilot channel through the Elephant Butte delta; aiding the federal government in maintaining the floodway of the Middle Rio Grande project; constructing and evaluating the success of in-river habitat for the Rio Grande silvery minnow; constructing hatcheries for the silvery minnow; conducting hydrologic investigations; developing and using numerical models to better understand and manage the river system; and serving as technical resource for the silvery minnow litigation and the litigation threatened by Texas on the Lower Rio Grande several years ago. Prior to working for the ISC, Mr. Schmidt-Petersen worked as a hydrogeologist for Daniel B. Stephens &

Associates, Inc. He has seventeen years of experience working in New Mexico on hydrology related issues. Mr. Schmidt-Petersen graduated from the New Mexico Institute of Mining & Technology with an M.S. in hydrology.

Scholle, Peter A.
Director and State Geologist
New Mexico Bureau of Geology and Mineral Resources
New Mexico Institute of Mining and Technology
801 Leroy Place
Socorro, NM 87801
(505) 835-5294
scholle1@nmt.edu

Peter Scholle is the director and state geologist for the New Mexico Bureau of Geology and Mineral Resources. Peter has had a rich and diverse career in geology: nine years with the U.S. Geological Survey, four years directly employed by oil companies (plus many additional years of petroleum consulting), seventeen years of teaching at two universities, and now a career in New Mexico state government. His main areas of specialization are carbonate sedimentology and diagenesis as well as exploration for hydrocarbons in carbonate rocks throughout the world. He has worked on projects in nearly twenty countries, with major recent efforts in Greenland, New Zealand, Greece, Qatar, and the Danish and Norwegian areas of the North Sea. A major focus of his studies dealt with understanding the problems of deposition and diagenesis of chalks, a unique group of carbonate rocks that took on great interest after giant oil and gas discoveries in the North Sea. His career has also concentrated on synthesis of sedimentologic knowledge with the publication of several books on carbonate and clastic depositional models and petrographic fabrics. He and his wife have published many CD-ROMs for geology, oceanography, and environmental science instructors. He has been president of the Society for Sedimentary Geology and treasurer of the American Geological Institute. He is currently past president of the Association of American State Geologists. Peter Scholle received a B.S. in geology from Yale University and continued his studies at the University of Munich on Fulbright/DAAD Fellowships and at the University of Texas at Austin. Scholle received M.S. and Ph.D. degrees in geology from Princeton University.

Skaggs, Rhonda K.
Dept. of Agricultural Economics & Agricultural Business
New Mexico State University
Box 30003, MSC 3169
Las Cruces, NM 88003
(505) 646-2401 Fax: (505) 646-3808
rskaggs@nmsu.edu

Rhonda Skaggs is a professor of Agricultural Economics and Agricultural Business at New Mexico State University. Her teaching and research interests include agricultural policy, agricultural structure and natural resource management, the future of agriculture, irrigation economics, and U.S.–Mexico cattle industry issues.

Titus, Frank
2864 Tramway Circle NE
Albuquerque, NM 87122-2289
(505) 856-6134
aquagadfly@aol.com

Frank Titus arrived in Albuquerque 51 years ago to work for the U.S. Geological Survey. In the next 37 years he was scientist and educator for the USGS and the New Mexico Institute of Mining and Technology, manager of Environmental Impact Statements across the U.S. and Canada, then back to Albuquerque as hydrology manager on the federal Uranium Mill Tailings Remedial Action Project. Titus says he "selectively retired" in 1993. Thereafter, his career focus has been on wise management of water and environment. In this role he has been: science advisor to State Engineer Tom Turney; facilitator on three decision-maker field conferences; in several TV specials on Rio Grande water; on the Middle Rio Grande Water Assembly; and a member of numerous advisory committees and boards of directors of non-governmental organizations, including the Water Dialogue. He has published numerous Op Ed columns in regional newspapers, co-authored a booklet entitled *Taking Charge of Our Water Destiny*, lectured on water affairs to numerous public groups, and often testified on water before committees of the state legislature.

Towne, Leann
Manager, Water Management Division
Albuquerque Area Office
U.S. Bureau of Reclamation
555 Broadway NE, Suite 100
Albuquerque, NM 87102
(505) 462-3579 or (505) 235-3616
ltowne@uc.usbr.gov

Since 1999 Leann has worked in the Bureau of Reclamation's Albuquerque Area Office, which has responsibility for projects on the Rio Grande, and Pecos and Canadian rivers. She is currently the Water Management Division Manager with responsibility for daily water operations and long-term water management for the Upper Rio Grande Basin above Elephant Butte Reservoir, the Pecos River Basin in New Mexico, and various other projects and programs. She deals with a number of water management issues that have significant implications for New Mexico, including meeting endangered species needs and delivery of water to senior users, such as the pueblos. Leann has worked for the U.S. Bureau of Reclamation since graduation, beginning her career in Durango, Colorado. During more than sixteen years of service with the Bureau of Reclamation, she has worked on various aspects of numerous projects, including water operations and management, facility maintenance and rehabilitation, and safety of dams. Leann grew up in New Mexico and graduated from New Mexico State University with a B.S. in civil engineering.

CREDITS FOR PHOTOS AND GRAPHICS

Unless otherwise noted, all graphics in this volume were completed by bureau staff from material provided by the authors. Individual photographers hold copyright to their works, which are reproduced here with permission.

Front cover, i	Adriel Heisey
5, 6	William Stone
8	Courtesy Palace of the Governors (MNA/DCA), # 39350
11L	R. L. Chapman, courtesy of the U.S. Geological Survey
11R	William L. Graf
18	U.S. Army Corps of Engineers photo (1963)
23	1947: Soil Conservation Service (NRCS); 1959 & 1966: U.S. Geological Survey. All three photos courtesy of the Earth Data Analysis Center photo archives, University of New Mexico.
35, 36	Adriel Heisey
46, 47, 48	Courtesy of the authors
51	Original hydrograph provided by Dr. Chris Young.
52, 53	Bosque del Apache National Wildlife Refuge
58, 59	New Mexico Interstate Stream Commission
60	Pete Balleau
63, 64	Adriel Heisey
66	Information was provided by Viola Sanchez of the U.S. Bureau of Reclamation, Kevin Flanigan of the New Mexico Interstate Stream Commission, and Robert Gold of the U.S. Geological Survey.
69	Adriel Heisey
77, 78, 79	Bosque del Apache National Wildlife Refuge
80	Jim Bones
81	Suzanne Landridge, U.S. Geological Survey
82L	Bosque del Apache National Wildlife Refuge
82R	New Mexico Interstate Stream Commission
85, 86, 87	Bosque del Apache National Wildlife Refuge
90	Adapted from an illustration provided by Nabil G. Shafike, New Mexico Interstate Stream Commission.
91	Adapted from an illustration provided by Vince Tidwell at Sandia National Laboratories.
95, 96	Adriel Heisey
98, 99R, 100	Office of the State Engineer
99L	Bureau of Reclamation
Back cover:	Clockwise from top left: William Stone, Adriel Heisey, Jim Bones, Adriel Heisey, Adriel Heisey, James McCord.

ACRONYMS

AWRM	Active Water Resource Management
CALFED	California Bay–Delta Ecosystem Restoration Program
CREATE	Center for Rapid Environmental Assessment and Terrain Evaluation
EDAC	Earth Data Analysis Center
EPSCoR	Experimental Program to Stimulate Competitive Research
ESA	Endangered Species Act
ET	evapotranspiration
FWS	U.S. Fish and Wildlife Service
GIS	Geographic Information System
HIA	historically irrigated acres
ISC	Interstate Stream Commission
LFCC	Low Flow Conveyance Channel
LTER	Long Term Ecological Research Center
MRG	Middle Rio Grande
MRGCD	Middle Rio Grande Conservancy District
NASA	National Aeronautics and Space Administration
NMISC	New Mexico Interstate Stream Commission
NWR	National Wildlife Refuge
OSE	Office of the State Engineer
PACE	Purchase of Agricultural Conservation Easement
PIA	practically irrigable acreage
PIRG	Public Interest Research Group
SWCD	Socorro Soil and Water Conservation District
URGWOM	Upper Rio Grande Water Operations Model
WATERS	Water Administration Technical Engineering Resource System
WRRI	Water Resources Research Institute

GENERALIZED GEOLOGIC MAP

The Rio Grande here follows the axis of the Rio Grande rift, one of the most striking geologic features in New Mexico. The rift is a complex of down-faulted blocks of crystalline and pre-Tertiary sedimentary rocks overlain by sedimentary basin fill. Rifting began here approximately 30 million years ago. Tributary drainages have contributed much of the alluvium that has been reworked into the floodplain of the Rio Grande. It is these younger Quaternary alluvial sediments that make up the shallow aquifer in the floodplain. Outside the floodplain the regional aquifer is primarily alluvial deposits of the (older) Santa Fe Group. The Santa Fe Group in this region ranges in thickness from 1,500 to 6,500 feet.

The uplifted flanks of the rift are composed of cores of Proterozoic basement overlain by sediments of Paleozoic and Mesozoic age. Paleozoic sediments of Pennsylvanian and Permian age are well exposed in the San Andres Mountains to the southeast and in the Los Pinos Mountains just east of Sevilleta National Wildlife Refuge. The mountain ranges to the west (including both the San Mateo Mountains and the Magdalena Mountains) contain extensive deposits of mid-Tertiary volcanic rocks, remnants of explosive volcanic activity that occurred throughout much of southwest New Mexico from 35 to 20 million years ago. Many of the lower-elevation areas on the map are covered by Quaternary alluvial or eolian (windblown) sediments. The large patch of Quaternary volcanic rocks on the floor of the Jornada del Muerto southeast of Fort Craig are young (780,000-year-old) basalts. The predominantly north/south orientation of the mountain ranges on both sides of the rift reflects the east-west extension of the crust that is characteristic of the Rio Grande rift.

Elephant Butte Reservoir is shown on the geologic map near its full capacity, extending to the northern boundary of Elephant Butte State Park near Fort Craig. Today the reservoir is considerably smaller in extent, closer to what is shown on the topographic map. The Low Flow Conveyance Channel (on the west side of the Rio Grande between San Acacia and Elephant Butte) is shown in red on both maps. The infrastructure along this stretch of river is shown in more detail on maps on pages 28 and 38.

This map was designed to accompany the full-page map on page viii of this volume and includes the Middle Rio Grande from the Socorro/Valencia County line to Elephant Butte Reservoir. It was excerpted by Kathy Glesener and Glen Jones from the Geologic Map of New Mexico (1:500,000), published in 2003 by the New Mexico Bureau of Geology and Mineral Resources. The digital elevation model was created from 10-meter data and provides a topographic overlay for the geologic map, as well. The complete state geologic map is available for purchase from the New Mexico Bureau of Geology and Mineral Resources. Call (505-835-5490) or visit our Web site at www.geoinfo.nmt.edu to order.

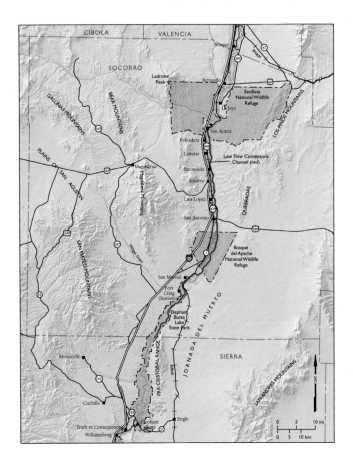

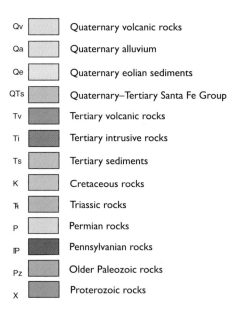